INFORMATICS AND THE DIGITAL SOCIETY

Social, Ethical and Cognitive Issues

IFIP - The International Federation for Information Processing

IFIP was founded in 1960 under the auspices of UNESCO, following the First World Computer Congress held in Paris the previous year. An umbrella organization for societies working in information processing, IFIP's aim is two-fold: to support information processing within its member countries and to encourage technology transfer to developing nations. As its mission statement clearly states,

IFIP's mission is to be the leading, truly international, apolitical organization which encourages and assists in the development, exploitation and application of information technology for the benefit of all people.

IFIP is a non-profitmaking organization, run almost solely by 2500 volunteers. It operates through a number of technical committees, which organize events and publications. IFIP's events range from an international congress to local seminars, but the most important are:

- The IFIP World Computer Congress, held every second year;
- open conferences;
- working conferences.

The flagship event is the IFIP World Computer Congress, at which both invited and contributed papers are presented. Contributed papers are rigorously refereed and the rejection rate is high.

As with the Congress, participation in the open conferences is open to all and papers may be invited or submitted. Again, submitted papers are stringently refereed.

The working conferences are structured differently. They are usually run by a working group and attendance is small and by invitation only. Their purpose is to create an atmosphere conducive to innovation and development. Refereeing is less rigorous and papers are subjected to extensive group discussion.

Publications arising from IFIP events vary. The papers presented at the IFIP World Computer Congress and at open conferences are published as conference proceedings, while the results of the working conferences are often published as collections of selected and edited papers.

Any national society whose primary activity is in information may apply to become a full member of IFIP, although full membership is restricted to one society per country. Full members are entitled to vote at the annual General Assembly, National societies preferring a less committed involvement may apply for associate or corresponding membership. Associate members enjoy the same benefits as full members, but without voting rights. Corresponding members are not represented in IFIP bodies. Affiliated membership is open to non-national societies, and individual and honorary membership schemes are also offered.

INFORMATICS AND THE DIGITAL SOCIETY

Social, Ethical and Cognitive Issues

IFIP TC3 / WG3.1&3.2 Open Conference on
Social, Ethical and Cognitive Issues of Informatics and ICT
July 22–26, 2002, Dortmund, Germany

Edited by

Tom J. van Weert
Hogeschool van Utrecht
The Netherlands

Robert K. Munro
University of Strathclyde
Scotland

KLUWER ACADEMIC PUBLISHERS
BOSTON / DORDRECHT / LONDON

Distributors for North, Central and South America:
Kluwer Academic Publishers
101 Philip Drive
Assinippi Park
Norwell, Massachusetts 02061 USA
Telephone (781) 871-6600
Fax (781) 681-9045
E-Mail <kluwer@wkap.com>

Distributors for all other countries:
Kluwer Academic Publishers Group
Post Office Box 322
3300 AH Dordrecht, THE NETHERLANDS
Telephone 31 78 6576 000
Fax 31 78 6576 254
E-Mail <services@wkap.nl>

Electronic Services <http://www.wkap.nl>

Library of Congress Cataloging-in-Publication Data

A C.I.P. Catalogue record for this book is available from the Library of Congress.

Informatics and the Digital Society: Social, Ethical and Cognitive Issues
Edited by Tom J. van Weert and Robert K. Munro
ISBN 1-4020-7363-1

Printed on acid-free paper.
Printed in Great Britain by IBT Global, London

Contents

Acknowledgements

SECIII would not have been the tremendous success it proved to be without the help and support of many, many people.

Without Sigrid Schubert there would have been no SECIII. She had the vision and worked tirelessly, often under very difficult personal circumstances, to ensure the vision was achieved. IFIP, the International Programme Committee and the National Organising Committee also invested a tremendous amount of time and effort over a couple of years to bring SECIII to fruition. They were helped by the many sponsors of the conference - Sun Microsystems, the dortmund-project, Stadt Dortmund, IBM, Siemens, ITC Dortmund, Apple and the Fachhochschule Dortmund. The University of Dortmund provided excellent facilities and hosted the event immaculately.

This book encapsulates the work of authors from around the globe. The editors would like to thank these authors most sincerely for the care and thought that went into all their papers and the extremely friendly cooperation which infused the editing process. Without exception the authors met deadlines, provided additional diagrams when requested and were always ready to reappraise and modify their texts. This friendly cooperation made the editors' job so much easier.

Unfortunately we were unable to include every SECIII presentation but we were able to devote one section of this book to Working Group Reports. While all the group participants must be commended for their involvement and interaction throughout the group discussion process we'd like to offer particular thanks to the rapporteurs who supplied us with such comprehensible summary reports.

Sigrid Schubert turned her vision into reality. Some colleagues deserve a special mention for turning the collective vision of the authors' papers into the reality of this book. Deborah Doherty at the Kluwer "dthelp" desk managed to resolve major technical problems so willingly, calmly and effectively when one editor thought all was lost. Finally, Yana Lambert, the IFIP editor at Kluwer Academic Publishers, was extremely supportive of the editors throughout the editing/publishing process and proved incredibly helpful (and so very patient) in ensuring all the many text and graphic flaws were ironed out of our last-gasp, last-draft manuscript. We couldn't have done it without you!

Preface

SECIII – Social, Ethical and Cognitive Issues of Informatics and ICT

Welcome to the post-conference book of SECIII, the IFIP Open Conference on Social, Ethical and Cognitive Issues of Informatics and ICT (Information and Communication Technology) which took place from July 22-26, 2002 at the University of Dortmund, Germany, in co-operation with the German computer society (Gesellschaft für Informatik). Unlike most international conferences, those organised within the IFIP education community are active events. This wasn't a dry academic conference – teachers, lecturers and curriculum experts, policy makers, researchers and manufacturers mingled and worked together to explore, reflect and discuss social, ethical and cognitive issues. The added value lies in what they, the participants, took away in new ideas for future research and practice, and in the new networks that were formed, both virtual and real.

In addition to Keynote Addresses and Paper Presentations from international authors, there were Provocative Paper sessions, Case Studies, Focussed Debates and Creative Exchange sessions as well as professional Working Groups who debated particular themes. The Focussed Debate sessions helped to stimulate the sense of engagement among conference participants. A Market Place with follow-up Working Groups was a positive highlight and galvanised participants to produce interesting reports. These were presented to the conference on its last day. Cross-fertilisation between the papers generated some surprising and useful cross-referencing and a plethora of social, ethical and cognitive issues emerged in the discussions that followed the paper presentations.

As the Patroness of SECIII, the Federal Minister of Education and Research, Edelgard Bulmahn, commented: "Information and communication technologies (ICT) increasingly enter all industrial and service sectors – not only in Germany, but world wide. The rapid development has led to a large supply of highly qualified jobs and a shortage of skilled manpower. We are all facing similar challenges which are reflected in the concept of this meeting: How can we support young people's, and especially young women's, interest in studies and vocational training in the ICT professions? How can we create reliable systems providing effective protection against misuse, thus building trust? What contributions can collaboration between educational institutions and industry make? How can we make sure that the potential of information and communication technologies will benefit all; how can we prevent a divide between an 'information elite' on the one hand and the less informed people on the other?"

As mentioned SECIII was an interactive conference in which these and many other questions and issues were raised and discussed. For example, in the Provocative Paper sessions the following issues were debated:

- One notebook per teacher: A sustainable concept for a wide ICT integration in school, Beat Döbeli Honegger and Marc Pilloud;
- Computer science as a profession in Germany: An academic perspective, Esther Ruiz Ben and Britta Schinzel;
- Practical use of multi-media in teacher training, Friedhelm Schumacher;
- Ethical apprenticeships?, Barry Blakeley;
- Paradigm shift on education through thematic modelling under object oriented support, Clara Amélia de Olivera.

In the Market Place conference participants presented views, materials and ideas on whiteboards for discussion with interested colleagues. This provided input for various Working Groups that were active during the conference. On the final day of the conference each Working Group gave a short presentation on the results of their work. All of the reports of the Working Groups have been included in this book. The results from the groups show strong convergence with the issues to be discussed at the forthcoming World Summit on the Information Society (WSIS Geneva 2003) [http://www.itu.int/wsis/]

The SECIII conference was held on the large, modern campus of the University of Dortmund. The university has about 25 thousand students and offers more than 30 Masters programmes, from mathematics to journalism. Teacher education is offered at all levels. With 3,500 students the Department of Computer Science is Germany's largest Computer Science department. It offers theoretical and applied programmes as well as teacher education programmes in Computer Science.

The campus of the University of Dortmund is located close to the successful Technology Park where the 'Dortmund-Project' helps to set up new anchor industries in information and communication technologies, micro-system engineering and e-logistics. With the disappearance of heavy industry from Dortmund, research and technological innovation are the new raw materials of the region's economy.

Conference Organisation

That this conference attracted such a large number of interested academics and ran so smoothly is a credit to all of the people involved in organising SECIII - IFIP, the International Programme Committee and the National Organising Committee.

The IFIP Contribution

Two IFIP Working Groups adopted the organisation of the programme of SECIII:

- Working Group 3.1 on Secondary Education, Chair: Deryn Watson,
- Working Group 3.2 on Higher Education, Chair Tom van Weert (past)/ Joe Turner (present).

More information on IFIP and its Working Groups can be found on: http://www.ifip.or.at

International Programme Committee

Yvonne Bütner (Basel, CH)
Volker Claus (Stuttgart, D
Hélène Godinet (Lyon, F)
Raymond Morel (Geneva, CH)
Robert Munro (Glasgow, UK)
Sigrid Schubert (Dortmund, D)
Deryn Watson (London, UK), Chair
Tom van Weert (Utrecht, NL), Vice Chair
Raul Wazlawick (Santa Catarina, BR)

National Organising Committee

Torsten Brinda (Dortmund)
Volker Claus (Stuttgart)
Hans Decker (Dortmund)
Hans-Ulrich Dönhoff (Düsseldorf)
Stefanie Gerszewski (Dortmund)

Burkhard Igel (Essen)
Volker Kampmeier (Soest)
Marie Kuhl (Dortmund)
Martin Lutz (Dortmund)
Johannes Magenheim (Paderborn)
Athanasios Papoulias (Dortmund)
Sigrid Schubert (Dortmund), Chair
Birgit Sirocic (Dortmund)
Wolfgang Weber (Soest)

The Conference Themes

Within the broad title for the SECIII conference four themes were identified as being of critical importance. The keynote presentations, contributed papers and working group debates all focused on these themes.

The e-literate society – the role of informatics, computer science and ICT

Everybody should be taught, should know and should understand key principles and acquire a range of transferable skills, which would help them to use the technologies of today as well as the technologies of the future. This would help to resolve many of the equity issues which exist in society, e.g. access to technology and information, equal opportunities for men and women, disadvantaged groups, loss of cultural richness and diversity. What role could the educational system play in determining the knowledge and skills and creating a more equitable society? What criteria should be developed to measure and evaluate success in this field?

ICT – agent of change and social conflict

The rapid development of ICT has led to many conflicts in society and sharpened specific issues: unemployment, differential access to information, cyber crime, the generation gap, North-South divide. On the other hand it offers promising perspectives: ICT has dramatically transformed our daily life and work, new professions are emerging, new economic frameworks are developing. New educational initiatives, such as life long learning and multimedia, have been formulated. Should education be making students aware of these issues by exploring ways in which these could be resolved? How can education increase social responsibility and integrate it into learning? Will society have to accept crime (e.g. on the Internet), social disorder and disadvantage and how can education of computer scientists produce socially responsible persons? Can the agent of change and social conflict be harnessed to resolve these problems?

E-learning – meeting the challenge of technology on society through new partnerships

Educators, teachers, trainers and their students are expected to use ICT in more imaginative ways in all aspects of teaching and learning. In facing up to this, new partnerships are being established e.g. between schools and industry (in creating courseware and teaching modules), between different educational institutions (in developing online-courses) or between local social groups to engage persons currently outside the computer community. In all cases e-learning offers new experiences to education and poses the challenge of developing new tools like authoring systems, intelligent tutor agents, and co-operative and collaborative learning environments that shall realise the potential of ICT.

Paradigm shifts in education and professional life

New technologies offer new possibilities for teaching. Education must develop a new paradigm to accommodate the technologies and these possibilities. It's important to integrate the impact of technology on society, but generally educators are not well prepared or motivated to explore issues associated with law, ethics or media education. What tools, what content, what experiences and what resources should be designed to help shape the new paradigm of didactics of computing and what concepts should be adopted?

The Conference Proceedings

The structure of this book was relatively obvious. The Keynote Sessions were all about generating and stimulating debate that lasted throughout SECIII. Without exception the Keynotes were extremely successful, sparking discussion - even argument - and forging partnerships and small groups who explored controversial issues in great detail and then shared their conclusions readily with all conference attendees.

We decided that the first section of the book should present all of the Keynote papers and that it would be valuable to present the international papers associated with each of the four themes, organised into their sub-topics. The final section of the book should present the reports from the Professional Working Groups. Importantly these Reports provide paths down which development of issues might proceed or ways in which specific problem issues might be best addressed.

SECIII Conference Proceedings therefore takes the following form:

Stimulating debate - themes and keynotes

Two keynotes address *The e-literate society - the role of informatics, computer science and ICT*. These are:

- Key Issues in IFIP-SIG9.2.2 Approaches to Ethics of Computing by Jacques Berleur, and
- Informatics - The Science of Minimal Systems with Maximal Capacity by Andreas Schwill

The theme of *ICT - agent of change and social conflict* was addressed in the keynote:

- ICT in Education: Aspirations and Tensions by David Wood.

The theme of *E-learning – meeting the challenge of technology on society through new partnerships* is explored in:

- E-learning Technology: Convergence with the Mainstream by Colin Harrison.

Finally two keynotes focus on the theme of *Paradigm shifts in education and professional life*:

- Knowledge Management in Education by Jane Andersen and
- Learning, and Teaching in Socio-technical Environments by Thomas Hermann.

Working Group Reports

This section includes all reports produced by international working groups during the conference:

E-literacy,
Collaborative Learning,
Social Issues and the Shift of Power,
Virtual Reality,
Equity and the Digital Divide,
Imported Culture,
Cyber Risk, Security and Privacy,
ICT Learning Environments,
E-Learning.

The e-literate society-the role of informatics, computer science and ICT

Three papers deal with the topic of Ethics of ICT in society.

The issues involved in Informatics Teaching were very popular and this section includes eight papers on that topic.

Finally there are two quite different papers which address the area of Special Education.

ICT - agent of change and social conflict

Associated with this theme are papers on Virtual Communities, on Collaborative Learning and on Social Issues and the Shift of Power.

E-learning – meeting the challenge of technology on society through new partnerships

The papers on this theme deal with innovative pedagogical practices and e-learning in the traditional university.

Paradigm shifts in education and professional life.

The paper here deals with teacher education in the interplay of ICT and society.

The Editors

Tom J. van Weert holds the chair in ICT and Higher Education of the Hogeschool van Utrecht, University of Professional Education and Applied Science, The Netherlands. Earlier he was managing director of Cetis, Expert Centre for ICT-based Innovations in Higher Education of the same university. Before that he was director of the School of Informatics (Computing Science) of the Faculty of Mathematics and Informatics of the University of Nijmegen, The Netherlands. Tom has studied applied mathematics and computing science. He started his working career in teacher education and software engineering. He has been chair of the International Federation for Information Processing (IFIP) Working Groups on Secondary Education and Higher Education. He is an elected personal member of IFIP Technical Committee 3 on Education.

Robert Munro is a reader in the Department of Business and Computer Education at the University of Strathclyde in Scotland. He has been involved in using ICT in education for almost twenty years - helping teachers from diverse subjects to integrate ICT into their teaching and learning, designing and developing software for use in geography and history, and conducting research into uses of ICT and the ICT competences of teachers. He has directed major research projects in Scotland and in Europe and has a particular interest in the societal issues related to ICT use, in the identification of indicators of qualitative improvements in education resulting from the uses of ICT and in helping to foster e-learning.

Key Issues in IFIP-SIG9.2.2 Approaches to Ethics of Computing

Jacques Berleur
Institut d'Informatique, Facultés Universitaires Notre-Dame de la Paix, Namur, Belgium

jberleur@info.fundp.ac.be

Key words: Codes of Ethics/Conduct, Curricula, Educational Programmes, IFIP, Ethics, Ethical issues, Law, Self-regulation, Spaces for Discussion

Abstract: Within IFIP ethics has been a preoccupation for a long time. Debate started formally in 1988 about a proposed international 'Code of Ethics'. The 1992 General Assembly concluded that the time was not ripe to adopt such a Code and asked a TC9 Task Force to provide recommendations. First, a full analysis of the current codes of IFIP member societies was undertaken. This enlightened the major deontological professional issues, such as respectful general attitude, conscientiousness, competence, promotion of privacy and confidentiality, and transparency of information. A special interest group (SIG9.2.2) was established in 1994 to support the creation of 'spaces for discussion' where ethical debate could be promoted and supported throughout IFIP and other constituencies. In a second phase, SIG9.2.2 confronted ethics and the governance of the Internet. It published a monograph that developed the main Internet-related issues that had appeared in new charters, commandments, codes and guidelines. The third phase focuses on the relationship between ethics and self-regulation. Our conclusions are that matters of interest to business and commerce tend to have legal force and regulation, but the real principles and issues of ethics tend still to be subject to no legal force. Society needs to confront this.

1. INTRODUCTION

When speaking about computers and ethics everybody quotes James Moor:

"computer ethics is the analysis of the nature and social impact of computer technology and the corresponding formulation and justification of policies for the ethical use of such technology." (Moor 1985)

Without mentioning specific ethical issues, Moor insisted that computer ethics must involve technology and substantial scientific and technological evaluation. In his perspective it is difficult to make a clear distinction between computer ethics and computers and society issues. Other influential authors were Deborah Johnson (Johnson 1985; 1994; 2001) and Joseph Weizenbaum. The latter explicitly touched upon the question of ethics, indicating the types of research that he would not undertake, except after careful consideration and with defined limits (Weizenbaum 1976). Similar preoccupation was noted in the ACM Computing Classification Systems under 'Computing Profession' - but the word ethics does not appear explicitly before 1991. Amongst curriculum proposals, an ACM/IEEE-CS Task Force in 1991 proposed a specific module on 'Social, Ethical, and Professional Issues' covering topics like the historical and social context of computing, responsibilities of the computing professional, and intellectual property rights (ACM/IEEE-CS 1990). Other modules have been added in the 2001 ACM/IEEE-CS proposal (ACM/IEEE-CS 2001).

In this paper, I will initially consider the most recent proposals of curricula, which are the structure that specialists find suitable for the field of *Ethics of computing* (my preferred term). I will then confront, from experience that we have in IFIP since 1988, theory and practice, and derive recommendations for defining the most urgent ethical issues in computing today, and those considered worthy of a specific educational programme. I will outline the main themes that appear in textbooks and reflect current thinking. IFIP is an academic and professional body and my reflection will be mostly at that level.

2. RECENT STRUCTURED PROPOSALS IN CURRICULA FOR COMPUTER SCIENTISTS

The widely publicised report 'Consequences of Computing: A Framework for Teaching the Social and Ethical Impact of Computing' tries to link the topics of ethical analysis and the levels of social analysis (ImpactCS 1996).

The proposal exhibits certain weaknesses - the heterogeneity of some categories in the levels of social and the unclear distinction between responsibility and ethical issues (Berleur 1996). Nevertheless, their grid, which is proposed for different technologies such as Electronic Communication, Medical Technology and Artificial Intelligence, helps teachers and students remain open-minded about the main issues associated with specific uses of ICT. It allows the mapping of Levels of Social Analysis (individuals, communities and groups, organisations, cultures, instiutional sectors, nations and global) against a range of Topics of Ethical Analysis (responsibilities and ethical issues). The main idea is that in discussing problems people may acquire desirable social and ethical skills. However, the ethical issues which are covered: individual and professional responsibilities, and more specifically quality of life, use of power, risks and reliability, property rights, privacy, equity and access, honesty and deception, are in a way privileged. I recognise that most of those issues have an ethical content, but I do not know why those categories were chosen and not others.

Table 1. Social and professional issues

ITEM	TOPIC	HOURS
SP1	History of computing	1
SP2	Social context of computing	3
SP3	Methods and tools of analysis	2
SP4	Professional and ethical responsibilities	3
SP5	Risks/liabilities of computer-based systems	2
SP6	Intellectual property	3
SP7	Privacy and civil liberties	2
SP8	Computer crime	
SP9	Economic issues in computing	
SP10	Philosophical frameworks	

In 'Computing Curricula 2001', the Social and Professional Issues identified in 1991 have been considerably expanded, although they represent only 16 of the 280 core hours (ACM/IEEE-CS 2001). They are shown in Table 1. Even the last three categories, although very important, are elective hours.

These two curriculum examples illustrate the convergence of opinion about the key topics associated with the ethics of computing: professional and ethical responsibilities, risks and liabilities (and reliability), intellectual property, and privacy and civil liberties.

To complement this overview, Table 2 illustrates the topics covered in eleven Handbooks since 1985, Moor's publication date (Johnson 1985; 1995; 2001), (Ermann 1990), (Kallman 1993), (Forester 1994), (Oz 1994), (Johnson-Nissenbaum 1995), (Kizza 1998), (Spinello 2000), (Langford 2000).

Table 2. Books on ethics of computing

TOPIC	Included in HANDBOOKS
Privacy and security	10
Theory	9
Software ownership, Copyright, IPR	8
Liabilities (Defective programs)	7
Codes and Professional ethics	7
Legal issues	5
Computer crime	4
Hacking, Viruses	4
Power, Democracy	4
Quality of worklife	3
Justice (distribution of work) - Employment , Third World, AI	3
Whistle blowing	2
AI, Expert systems	2
Networked world	2
Free speech, Control of content	2
Internet Governance and regulation	1
Quality of personal life	1
Strategic Defence Initiative (SDI)	1
Risks of computing	1
Pornography	1
Encryption	1
Netiquette	1
Digital divide	1

If we consider the top categories, which appear in more than half of these books, it is obvious that the ethical issues which are perceived today as the most important may be summarised as follows (Table 3):

Table 3. Quoted ethical concerns

Broad Issue	Specific issue
Classical	Privacy
Classical	Risks and security
Classical	Intellectual property rights
Classical	Responsibility, reliability and liability in designing information systems or software
Classical	Professional codes of ethics or of conduct
Classical	Computer crime
Recent Internet-related	Equity of access
Recent Internet-related	Digital divide
Recent Internet-related	Pornography
Recent Internet-related	Netiquette

3. IFIP APPROACHES

3.1 Phase One: An IFIP Code for all member Societies or the creation of 'Spaces for discussion'?

What support did IFIP provide for structuring educational material in this emerging area? Officially the community started to debate ethics in the late 1980s, when its New Delhi General Assembly suggested the development of an international Code of Ethics. Even earlier, IFIP-TC9, and more precisely WG9.2, worked in this area. The question was also under discussion in the Council of Europe. After turbulent debate the idea of an international code was abandoned in favour of a process that would take more into account the cultural, social and legal diversity of the member Societies (Berleur 2001). Interestingly, although IFIP abandoned the idea more than 10 years ago, self-regulation today is curiously considered a solution to regulate internationally that which cannot be regulated by law because of its intrinsic link to territoriality.

Analysis of the codes of member Societies was carried out and suggestions were made, first to consider the results and other topics when drafting a code, and secondly, to create 'spaces for discussion' where deep convictions might be shared and from which principles could be derived (Habermas 1992; Maesschalck 1994).

No less than 15 IFIP member societies have codes (24 codes altogether). We added 6 other professional non-IFIP codes. The major ethical concerns these codes consider as belonging to the deontology of the profession comprise five main domains (Berleur and Gevers 1996).

Respectful general attitude

This includes: respect for the interests/rights of those involved [15/30], respect for the prestige of the profession [11], respect for the interests/rights of the public [10], and respect for the welfare/health of the public and for the quality of life [10].

Personal (/institutional) qualities: conscientiousness, honesty and positive attitude, competence and efficiency

In practice conscientiousness and honesty are frequently encountered under expressions like: acceptance of responsibility [19] and integrity [26], respect for requirements, contracts or agreements [14] and conscientious work [11]. With regard to competence and efficiency two other terms are very common: professional development and training [19] or limitation of work to the field of competence [18].

Promotion of information privacy and data integrity
Confidentiality [22] is required by nearly all the IFIP societies general codes [13/15]. Privacy in general [14] and respect for property rights [12] are appealed to often. No computer crime, no information piracy or misuse [7].

Production and flow of information
Flow of information to involved parties or people is required by the majority of the codes [23]. Information to the public [16] is also insisted upon. Almost half of codes calls for comprehensive information [14].

Attitude towards regulations
Regulations do not appear as a major theme. Less than half the codes require respect for the code [13], respect for the law [13], and respect for IT and professional standards [12]. Some consider sanctions against a breach of the code [9].

Clearly these codes contain general clauses and recommendations, which are not deeply rooted in ICT. Most clauses belong to general business ethics, except perhaps those related to privacy, data integrity, production and flow of information. Even these are the translation in the ICT field of the 'fair relationship with the client' in the field of business ethics.

IFIP recommended its member Societies should consider more ICT-related topics - issues such as computer crime, the accountability and liability for the content of networked information and transborder intellectual property rights. It was also suggested they should study in depth some issues of much broader scope and see how they might be included in codes of conduct. These covered the unequal distribution of information; unequal access to technology; participation in decisions which affect life at home and work; how computer technology reinforces predominant power; denial or restriction of access to groups/individuals who lack the resources to participate in an increasingly market dominated system; poor cultural diversity which today pervades our information, media and communication systems; the necessity of majoring local cultural content requirements in network offerings; the importance of promoting culturally sensitive and multi-cultural interfaces; and involving cultural minorities in the design of socially significant information and communication systems (IFIP GA 1996).

As a result of this first phase, IFIP's General Assembly recommended the creation of 'spaces for discussion' in each member Society and the establishment of a permanent Special Interest Group (IFIP-SIG9.2.2) on Ethics of Computing. Their objectives were to keep alive the ethical question within the different IFIP constituencies and in all circles where IFIP members manifest their influence. They were asked to suggest themes and topics to be studied nationally or regionally.

3.2 Phase Two: Ethics and the Governance of the Internet

After examination of IFIP member Societies codes, SIG9.2.2 analysed similar documents published 'outside its walls'. Many employed the old medieval term 'Charter', stressing their foundational character: 'Cyberspace and the American Dream: A Magna Carta for the Knowledge Age' (1994, Progress and Freedom Foundation, PFF); 'Proposal for an Internet charter: Rules and Courtesies of the Actors of the Internet' (France, 1997); *Online Magna Charta*, Charta of Freedom for Information and Communication', (1997, The Wartburg Charta); and the 'Intergovernmental Information Technology Leadership Consortium Charter' (1997, Council for Excellence in Government). Others, like 'Ten Commandments of the Computer Ethics Institute', used the majestic style of worship: 'Thou Shalt Not Use A Computer To Harm Other People. Thou Shalt Not Snoop Around In Other People's Computer Files'. The Internet Service providers associations were drafting their first codes (EuroISPA 1997; IAPCode 2001), and it was the time for a new approach to the problem of governing the Internet which spawned a fetish for the term 'governance'.

To record the analysis of these documents SIG9.2.2 adopted a grid commonly used for legal documents. This identifies the actors and the people concerned (*ratione personae*), the place where the law is applicable (*ratione loci*), the matters which are covered (*ratione materiae*), the rules for enforcement, including sanctions and procedures, and finally a 'miscellaneous' section.

Topics derived from analyses and which were considered ethical topics included (Rasmussen 2001):

- no illegal, dubious, harmful material, racial hatred, drug traffic, denial of crime against humanity;
- protection of children and of youth;
- confidentiality, privacy and other linked right such as access, correction of personal data;
- correctness, accuracy and fairness;
- respect, honesty, competence, sincerity, right information, work quality;
- no computer crime;
- fair trading, correct pricing information, consumer protection, consumer confidence, no unsolicited email advertising, reliable information, transparency, clear definition of liabilities;
- protection of intellectual property rights, copyright, trademark, patent;
- free speech, right to information and communication.

Many of these topics were already mentioned earlier (Table 3).

These results were published and discussed during the 1998 IFIP-TC9 Fifth Human Choice and Computers international conference in Geneva and presented to the 1999 IFIP General Assembly in Kuala Lumpur through a monograph that was distributed to all the member Societies to promote discussion.

3.3 Phase Three: Ethics and Self-regulation

Between 1998 and 2001, IFIP-SIG9.2.2 collected nearly 40 additional self-regulation documents. These have been analysed according to the same analysis grid described above and the results have been cumulated with those from the analysis of the 30 IFIP and professional codes. This enabled the construction of a first inventory: Table 5 offers a detailed classification of all 70 documents (Berleur and Ewbank 2002).

Table 4. Classification of the inventory of self-regulation documents

Numerical Class	Category
1	**Informatics- Computing in general**
1.1	General principles
1.2	Specific principles
1.2.1	For Professional Societies
2	**The Internet**
2.1	General principles
2.1	Specific principles
2.2.1	For Virtual Communities
2.2.2	For Different Actors
2.2.3	For (Associations of) Service Providers
2.2.4	For other Service Providers
2.2.5	For Governmental Services and Actors
2.2.6	For the Industry in general
2.2.7	For Specific Sectors and Services
2.2.7.1	Health Sector
2.2.7.2	Publishing Sector
2.2.7.3	e-Commerce Sector
2.2.7.4	Software Publishers Sector
2.2.7.5	Telemarketing Sector

While 'general principles' codes remain rather fuzzy they do have a deontological content. They may include consideration of correctness as in the rules of Netiquette. In inviting the respect of others and other people's work they could be considered to belong to the tradition of professional ethics or fair conduct. But they are not so coercive. The virtual communities and service providers express concerns about 'illegal material' (child pornography and racist propaganda), the necessity of protecting youth, (especially against those who exploit their credulity) and manifest their commitment to cooperate with hotlines. They also mention data protection, confidentiality and e-mail secrecy and include clauses on no violence, hatred, or cruelty, respect for human dignity and no discrimination be it ethnic/religious or handicap related.

Clauses related to 'computer crime' or 'cyber-crime' abound. They cover illegal access, illegal interception, data interference, system interference, misuse of devices, computer-related forgery and fraud, offences related to child pornography, infringements of copyright and related rights, attempt and aiding or abetting, as focused today by the recent Convention n° 185 on Cybercrime (Council of Europe 2001).

These are the more ethical topics. Curiously, with codes becoming more sectoral, one could/might expect more specific clauses, since self-regulation is supposed to anticipate or supplement what cannot be included, for instance, in law because of its territoriality or lack of flexibility, or the rapid development of technology. What clearly emerges is the more contractual character of these documents and the insistence on practice in accordance with the standards of the profession.

Deontology clauses become 'terms of service'. The Code of Ethics of the Internet Health Care coalition requires candour and trustworthiness, quality of information, products, services, the best *commercial* practices (*our emphasis*) and the highest standards by health care professionals (Internet Healthcare Coalition 2000). In online publishing, the French Charter reaffirms the normal professional rules but also details the number of paragraphs that can be quoted under fair dealing provisions and specifies the rules related to links that may be created to online content. In electronic commerce, the 'Model Code of Conduct for Electronic Commerce' offers typical clauses found in good commercial practice: reliability of the information which is provided, systems and organisations and types of electronic signatures, as well as transparency in the communication, confidentiality and respect for intellectual property (ECP-NL 2001).

If there are real ethical clauses, the impression is that when we leave the general principles of the deontological approach for more specific principles expressed in contractual terms, ethics increasingly is absent. While self-regulation instruments are supposed to protect all parties they appear to protect only the authors. Our deep conviction is that they are more self-protecting than self-regulating. The proliferation of codes of ethics and/or ethical conduct is not yet a sign of a 'rush towards ethics'.

If we want self-regulation to reflect real ethical concerns it must be scrutinised carefully. We have suggested that minimum requirements must be met and recommendations must be made on the level of self-regulation.

Codes of general principles are not the same as sectoral ones. The titles and preambles must make explicit the status of those documents, and they must all be negotiated in a participatory manner.

Self-regulation is becoming a major ethical problem and must be debated openly. While it is trying to find its place in the normative order it still remains minimal if not minimalist and most of the time, it is promulgated to avoid a greater degree of statutory regulation. Is self-regulation an anticipation, or a supplement or a substitute for the law? What is its ethical content? From a democratic point of view how legitimate is self-regulation? Who drafts and elaborates the legislation? What role should the regulators take to protect citizen and customers?

4. ETHICAL TOPICS FOR EDUCATIONAL PROGRAMMES

Patently many questions are interwoven. Codes of ethics or of conduct may include matters which are all relevant for ethics but are already covered by proper regulatory instruments. The Convention on Cybercrime, the European Directives on 'electronic communications networks and services' and the US Digital Millenium Copyright Act are excellent examples.

What is covered, among other matters is the protection of content (by protecting intellectual property and related rights); and the protection of the transaction (via the electronic signature for proper identification), the authentication of messages, and their confidentiality.

The Cybercrime Convention protects exchanges from intrusions and from illicit or fraudulent acts. The European Directive on electronic commerce and the US Digital Millenium Copyright Act free intermediaries from liability regarding content. Essentially, all that is linked to business and commerce is 'under control'.

This means that we need to focus our attention on topics with a more ethical content. SIG 9.2.2 made its choices on two levels. The first was rather pragmatic: the level of the questions which are already under discussion (or 'under control') including the protection of the individual (as citizen and as consumer), and the protection of organisations and of society.

These questions include ethical content, but there is less emergency. The second level is linked to principles. It establishes as principles of judgement and choice the protection of human dignity, the vulnerability of the weakest or the underprivileged, and the sustainability of society.

Among the different topics SIG 9.2.2 has met in its analyses the following issues involved more ethical content (Table 5):

Table 5. Topics with more ethical content

TOPIC
Equity in the right of access (universal service)
Questions linked to the respect of the dignity of the person
Justice and social inclusion (North-South, work distribution)
Respect for the interests and rights of the persons
Free speech/censorship
Quality of life
Right to information (transparency)
Personal qualities (honesty, competence)
Non-abuse of power (appropriate use)
Respect for cultural differences
Freedom of choice in the use or non-use of the Internet
Grounding 'virtual' life in the physical realm

Compare the first summary of the current ethical topics (Table 3) with that of the present list (Table 5).

Originally privacy, intellectual property rights, responsibility, reliability and liability in designing information systems or software, professional codes of ethics or of conduct, and computer crime were mentioned. I have explained that recent literature mentions specific issues, which arise from the development of the Internet: equity of access, the digital divide, pornography and Netiquette.

While there are some new topics, there is a great deal of convergence or correspondence between Table 3 and Table 5 - a convergence illustrated in Table 6.

The topics certainly supplement the list outlined in the report 'Consequences of Computing: A Framework for Teaching the Social and Ethical Impact of Computing' which was discussed early in this paper.

Philippe Quéau, Director of Information and Informatics at UNESCO, concluded his speech to the 1999 Regulators summit by remarking:

"The structural imbalances of the world-wide Internet infrastructure, the deep inequalities in the access to information, the transnational oligopolies controlling the world-wide infostructure are major preoccupations for the regulator. A new form of regulation or of governance at the world level must be conceived in an ethical world-wide perspective, at the service of equity and of the human development." (Quéau, 1999)

Educational programmes must prepare people to meet these requirements.

Table 6. Ethical issues - a summary

Most quoted ethical concerns (Table 3)	Topics with ethical content in self-regulation documents (Table 5)	Under control
Privacy		USA?
Risks and security		?
Intellectual property rights	Respect for interests and rights	Yes
Responsibility, reliability and liability in designing systems		?
Professional codes of ethics or conduct	Right to information (transparency) Personal qualities Non-abuse of power	
Computer crime		Yes
Equity of access	Equity in the right of access ('Universal Service')	Europe.Direct
Digital divide	Justice and social inclusion (North-South)	
Pornography	Questions linked to respect of dignity of the person (protection of minors and human dignity)	Self-regulation
Netiquette		?
	Free speech	?
	Quality of life	?
	Respect for cultural differences	?
	Freedom of choice re. Internet	?
	Grounding virtual life in the physical realm	?

5. A STEP ALONG THE WAY

Our approach may appear rather pragmatic but quite normal in applied ethics, where preconceived ideas should be avoided. The process must be rational and promote discussion on cultural, social and legal differences.

The content of the list proposed in the report 'Consequences of Computing: A Framework for Teaching the Social and Ethical Impact of Computing' is acceptable provided the levels of social analysis are slightly reformulated to be made more homogeneous, and provided the list of ethical issues is expanded and ranked in terms of priorities. The social and ethical skills are worthy of consideration for teaching. The social skills suggested by the ImpactCS report are:

"Identifying and interpreting the social context of a particular implementation; identifying assumptions and values embedded in a particular system; and evaluating, by use of empirical data, a particular implementation of a technology."

whilst the proposed ethical skills are:

"Arguing from example, analogy, and counter-example; identifying stakeholders in concrete situations; identifying ethical issues in concrete situations; applying ethical codes to concrete situations; and identifying and evaluating alternative courses of action." (ImpactCS 1996)

Our intention was to call for vigilance, to support education in 'technology watching', because the Internet induces and introduces new and major social and ethical changes. Looking back on thirty years of 'Computers and Society' challenges and fifteen years of 'Ethics and Computing' issues, we can observe similarities in the concerns, but the list of problems are not comparable. Indeed one list circulated by 'People for Internet Responsibility' identifies 116 categories covering 7 pages of documentation (PFIR 2001). In 1976 the IFIP General Assembly, establishing its Technical Committee 9 on 'Relationships between Computers and Society' judged that two Working Groups would be sufficient to cover the main questions!

However, I would like also to share my perplexity. I stated that specific laws have been enacted recently primarily to protect e-Commerce. Intellectual property and related rights was also recommended as a major ethical issue for consideration. What of questions related to human dignity and the protection of children, or about racial hatred?

With regard to the first, the European Union published a 'Green paper on the Protection of Minors and Human Dignity in Audiovisual and Information Services', in 1996. This was followed up by the 'Council Recommendation on the development of the competitiveness of the European audiovisual and information services industry by promoting national frameworks aimed at achieving a comparable and effective level of protection of minors and human dignity' in 1998. Next came a 'Safer Internet Action Plan' in 1999 where public authorities have almost no role to play. Recently, in 2002, the European Parliament issued a report on the evaluation of the Council Recommendation. This clearly stated that it does not intend to undertake any legislative measure on the subject.

Regarding racial hatred, the "Digital Hate 2002 Report" from the Simon Wiesenthal Center identified 600 extremist and racist sites 1997 and over 3400 in 2002 (Wiesenthal 2002). Sites feature animated hate games, online enrolment for suicide bombers, Neo-Nazi online identity theft and other examples of transnational hate. Extremism and racism pervade the domain of games such as the Austrian KZ (Konzentrationslager) Manager (where Turks are killed in a concentration camp), or the Swedish Alrea Alpy (where the targets in the shooting gallery are black people.)

Privacy is also an issue that, although present since the very first days of the computers and society discussions, has always been interpreted differently on the two sides of the Atlantic. Negotiation about the application of articles 25 and 26 of the 1995 European Directive on data protection, and the obligation for third parties to have an 'adequate level of protection' resulted in the famous 'Safe Harbor' self-regulation principles (COM 2000). Law on one side, self-regulation on the other!

Who will decide what is most important from the ethical point of view? We will - together. Faithful to our philosophy of opening spaces of discussion where we share our values and convictions, as I have with you. It will be up to each of us to do the same with others. The next agenda for IFIP-SIG9.2.2 will be the question of ICT, justice and social ex/inclusion.

REFERENCES

ACM/IEEE-CS (1990) *Computing Curricula 1991*. Report of the ACM/IEEE-CS Joint Curriculum Task Force, ACM Press, IEEE Computer Society Press, December. pp. 69-71.

ACM/IEEE-CS (2001) *Computing Curricula 2001*. Approved Final Draft of the Computer Science Volume, Joint Task Force on Computing Curricula, IEEE-CS/ACM, December. [http://computer.org/education/cc2001/]

Berleur, J. (1996) Remarks on a 'Framework for Social and Ethical Issues' Report. In *The Impact of Information Technology: From Practice to Curriculum*, Y. Katz, D. Millin and B.Offir (eds.), Chapman & Hall, pp. 108-117. See also: Same author, 'A Focus on the Main Issues...' Review of 'Computers in Context: A Framework for Presenting the Social and Ethical Impact of Computing. A Report from the ImpactCS Steering Committee'. In *Computers and Society*, ACM Special Interest group on Computers and Society, Volume 26, No. 1, March 1996, pp.7-8.

Berleur, J. and d'Udekem-Gevers, M. (1996) Codes of Ethics or of Conduct Within IFIP and Other Computer Societies. In *Ethics of Computing: Codes, Spaces for Discussion and Law*, J. Berleur and K. Brunnstein (eds.), Chapman & Hall (now Kluwer), London. pp. 3-41.

Berleur, J. and d'Udekem-Gevers, M. (2001) Codes of Ethics/Conduct for Computer Societies: The Experience of IFIP. In *Technology and Ethics, A European Quest for Responsible Engineering*, P. Goujon and B. Heriard Dubreuil (eds.). European Ethics Network, Peeters, Leuven, Belgium. pp. 327-350.

Berleur, J. and Ewbank de Wespin, T. (2002) Self-regulation: Content, Legitimacy and Efficiency - Governance and Ethics. In *Human Choice and Computers, Issues of Choice and Quality of Life in the Information Society*, K. Brunnstein and J. Berleur (eds.). Proceedings of the IFIP-HCC6 Conference, 17th World Computer Congress, Montréal,

August 2002, Kluwer Academic Publ. pp. 89-108. [http://www.info.fundp.ac.be/~jbl/IFIP/sig922/selfreg.html]

Commission Decision 2000/520/EC of 26.7.2000 pursuant to Directive 95/46/EC of the European Parliament and of the Council on the adequacy of the protection provided by the Safe Harbor privacy principles and related frequently asked questions issued by the US Department of Commerce - *Official Journal of the European Communities*, L 215/7 of 25.8.2000.

Council of Europe (2001) Convention on Cybercrime, Convention No.185, opened to signature on November 13, 2001.

[http://convention.coe.int]

Electronic Commerce Platform Nederland (2001) Code of Conduct for electronic commerce, (Draft 4.0, 2001)

[http://www.ecp.nl/ENGLISH/publication/cocdraft4.0ENG.pdf]

Ermann, M. D., Williams, M. B. and Gutierrez, C. (1990) *Computers, Ethics and Society,* Oxford University Press, New York, Oxford.

European Internet Services Providers Association (1997). Codes of Conduct of the Member Associations.

[http://www.eurooispa.org]

Forester, T. and Morrison, P.(1994) *Computer Ethics: Cautionary Tales and Ethical Dilemmas in Computing*, The MIT Press, 2nd edition.

Habermas, J. (1992) *De l'éthique de la discussion*, Paris, Cerf, 1992. [Orig.: *Erlauterungen zur Diskursethik,* Engl. Transl.: *Justification and Application: Remarks on Discourse Ethics,* Cambridge Mass.: The MIT Press, 1993].

IAPCode (2001) Selfregulation.info, Programme in Comparative Media Law and Policy, Centre for Socio-Legal Studies, Wolfson College, Oxford University. [http://www.selfregulation.info/]

IFIP GA (1996) Recommendations Regarding Codes of Ethics/Conduct for IFIP Computers and Society. IFIP General Assembly, Hamburg, September 6th, 1994, in: *Ethics of Computing: Codes, Spaces for Discussion and Law*, J. Berleur and K. Brunnstein (eds.). Chapman & Hall (now Kluwer), London. pp. 257-268.

Project ImpactCS Steering Committee (1996) 'The Consequences of Computing: A Framework for Teaching the Social and Ethical Impact of Computing'. In *Computers and Society, ACM SIGCAS*, vol. 26, No. 1. pp. 3-10. See also: Huff, C. and Martin, C. D. 'Computing Consequences: A Framework for Teaching Ethical Computing'. In *Communications of the ACM*, December 1995, vol. 38, Nr. 12, pp. 75-84.

(Internet Healthcare Coalition, 2000): Internet Healthcare Coalition, eHealth Code of Ethics. 2000.

[http://www.ihealthcoalition.org/ethics/]

Johnson, D. G. and Snapper J. W. (1995) *Ethical Issues In the Use of Computers*, Wadsworth Publishing Company. (Out of print).

Johnson, D. G. (1985; 1994; 2001) *Computer Ethics*, Englewood Cliffs, N. J.: Prentice-Hall, Inc.

Johnson, D. and Nissenbaum, H. (eds.) (1995) *Computers, Ethics and Social Values*, Prentice Hall.

Kallman, E. K. and Grillo, J. P. (1993) *Ethical Decision and Information Technology. An Introduction with Cases*, Mac Graw-Hill.

Kizza, J. M. (1998), *Ethical and Social Issues in the Information Age*, Springer.

Langford, D. (ed.) (2000) *Internet Ethics*, Palgrave - McMillan Press.

Maesschalck, M. (1994) *Pour une éthique des convictions. Religion et rationalisation du monde vécu.* Publications des Facultés universitaires Saint-Louis, Coll. Philosophie, Bruxelles.

Moor, J. H. (1985) What is computer ethics? In Computers and Ethics, W. T. Bynum (ed.). *Metaphilosophy*, Volume 16, No. 4. Basil Blackwell, Oxford and New York. pp. 266-275.

Oz, E. (1994) *Ethics for the Information Age*, Business and Educational Technologies, Wm. C. Brown Publishing, Dubuque, IA.

PFIR - People For Internet Responsibility (2001) ISSUES, Version of July 4. [http://pfir.org/issues]

Rasmussen, L. B., Beardon, C. and Munari, S. (eds.) (2001) *Computers and Networks in the Age of Globalization.* Proceedings of the 5th IFIP-TC9 'Human Choice and Computers' Conference, Kluwer Academic Publ., 2001: see the papers 'Ethics and the Governance of the Internet' and 'The Process of Ethics', pp. 141-154 and 361-379.

Spinello, R. (2000) *Cyberethics. Morality and Law in Cyberspace*, Jones and Bartlett Publishers, Sudbury Mass.

Quéau, P. (1999) 'Internet : Vers une régulation mondiale'. UNESCO, World Summit of regulators, Internet and the New Services, Paris.
[http://www.unesco.org/webworld/news/991201_queau_csa.shtml]

Weizenbaum, J. (1976) *Computer Power and Human Reason – From Judgment to Calculation,* W. H. Freeman and Coy. San Francisco. (Penguin Books edition in 1984, with a new Preface).

Simon Wiesenthal Center (2002) Press information, 20 May, 2002. [http://www.wiesenthal.com/social/press/pr_item.cfm?ItemID=5723]

Informatics - The Science of Minimal Systems with Maximal Complexity

Andreas Schwill
Institut für Informatik, Universität Potsdam, August-Bebel-Str. 89, D-14482 Potsdam, Germany

schwill@cs.uni-potsdam.de

Key words: Fundamental Ideas, Curriculum Research, Minimalism, Construction Kit

Abstract: It is a fundamental idea of computer science to search for, define, analyze, and operate with construction kits consisting of small sets of basic building blocks and a small number of operations to combine the building blocks to larger objects. While the construction kit is mostly simple, it often defines a vast, complex field that consists of all possible objects that can be built from the building blocks by using any (finite) sequence of combinations of operators. This idea affects and structures many areas of computer science. We present examples from several fields, including imperative and functional programming languages, computable functions, Turing and register machines, Boolean functions, data types, object-oriented programming, characterisations of formal languages along with examples from other disciplines. How can informatics lessons profit? If lessons are oriented towards a fundamental idea, the idea may explain, structure, and integrate many different informatics subjects and phenomena by a single recurring scheme. On the other hand, the construction kit principle belongs to the sphere of everyday thinking so students already have a basic intuition of the concept which may enhance their understanding when entering any field where the idea applies.

1. INTRODUCTION

In recent years we have elaborated Bruner's concept of fundamental ideas and made it accessible for informatics lessons (Bruner 1960). Here we consider in detail a fundamental idea of computer science - orthogonalization - and show that it has a wide area of application and may guide many fields of school informatics.

By *orthogonalization* of a field Δ, following a term in linear algebra, we denote the definition of a number of basic elements Δ_e of the field along with a set K of operations ($K=\{K_1,...,K_n\}$, n small) on the basis each as small and simple as possible, such that every other object of the field may be generated by finitely many applications of operations on the basic elements (Figure 1). The result is a minimal generating system $B=(\Delta_e,K)$, consisting of the basis and the operations, that may be considered as a construction kit for the field.

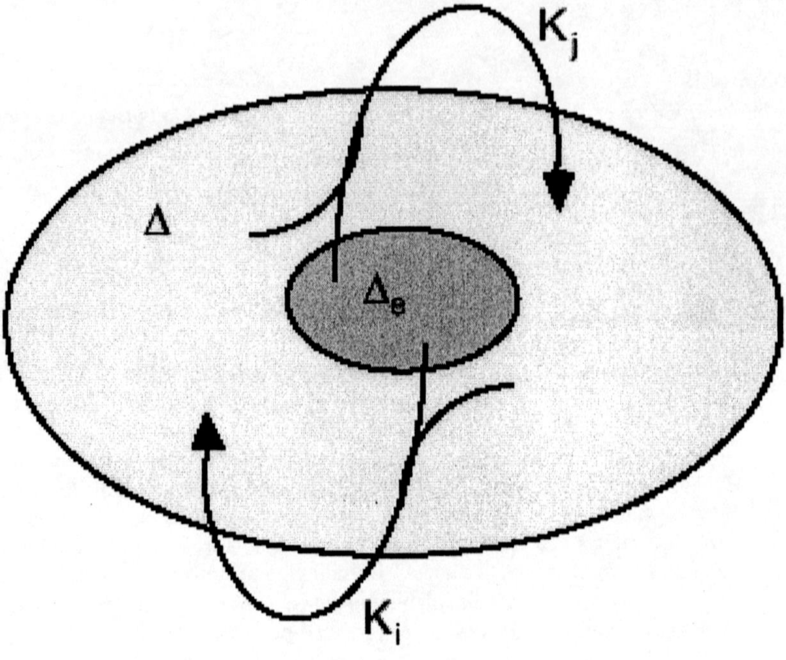

Figure 1. Principle of construction kits

We analyse orthogonalization with respect to didactic criteria and illustrate its relevance by presenting examples and applications in several areas of informatics.

We use the term *complex* to denote systems that are vast and diverse in their inner structure, while we call descriptions of systems *complicated* if they are vast and varied and hard to grasp. There is no direct relationship between the complexity of a system and the complication of its description. While the system, particularly a real life system, may be complex in nature, it may have a simple, short description. We must avoid complicated descriptions if the systems are simple, and search for descriptions as minimal as possible if the systems are complex, to be able to understand, master, or manage them.

2. BACKGROUND

In 1960 Bruner formulated the teaching principle that lessons should predominantly orient towards the structure (the so-called *fundamental ideas*) of science. In recent years we have adopted the concept, made it and the relevant notions precise, and transferred it to informatics lessons by defining fundamental ideas of informatics (including algorithmization, structural dissection, (artificial) languages and orthogonalization). We have also proposed lessons suitable for teaching certain ideas in school (Schwill 1993; Schwill 1997).

We define a fundamental idea as a schema for thinking, acting, describing or explaining which satisfies the following criteria:

Horizontal Criterion. A fundamental idea is applicable or observable in multiple ways and in different areas of informatics. It organizes and integrates a wealth of phenomena.

Vertical Criterion. A fundamental idea may be demonstrated and taught on every intellectual level - "any subject can be taught effectively in some intellectually honest form to any child at any stage of development" (Bruner 1960). A central methodological means guiding the education of fundamental ideas on different levels of understanding is the *spiral principle.* This recommends three representations of concepts to be learned - *enactive* (lower level), *iconic* (medium level), and *symbolic* (highest level).

Criterion of Time. A fundamental idea can be clearly observed in the historical development of computer science and will stay relevant in the future. Importantly lessons based on fundamental ideas will not become dated as quickly as conventional lessons - a major advantage in teaching informatics which exhibits such dynamic evolution.

Criterion of Sense. A fundamental idea has meaning in everyday life and is related to ordinary language and thinking. Only a precise definition turns an idea "with sense" into an exact concept "without sense". When we teach a fundamental idea early in the student's schooling, we may give a first impression of the idea by using everyday situations as starting points for lessons.

3. MINIMAL SYSTEMS IN ARTS, SCIENCES, AND INDUSTRY

Why do minimal systems in the form of construction kits play such an important role in the arts, sciences, and industry? There may be two main reasons. Firstly ecological/economical causality. Nature and industry develop evolutionarily to derive the maximum result by spending a minimum of resources. The other reason might be the limited capacity of the human brain, particularly short-term memory, so concepts should have small descriptions to be manageable by humans. Miller's finding that short-term memory capacity is 7±2 chunks (or self-contained objects) may explain why many minimal systems comprise less than 10 basic elements and operations (Miller 1956).

The following minimal systems found in arts, sciences, society, and industry also verify the Criterion of Sense for orthogonalization.

3.1 Orthogonalization in industry

A modern example is Volkswagen's platform strategy announced in 1997. By using only four platforms that contain 60% of the car's parts, including the chassis, engine, brake system and gearbox, Volkswagen wishes to eventually produce 51 different models. Further examples of orthogonalization occur in concepts of lean management or lean production.

3.2 Orthogonalization in society

Orthogonalization may be found in public management (Biedenkopf 1994). The complexity of problems in developed industrial societies permanently increases - in health, pensions, tax, or unemployment systems. At the same time the complication of methods, processes, and legal regulations grows to manage this increase in complexity. This accelerates the waste of financial, personal, and natural resources and produces little or no visible benefit to the public. Biedenkopf favours a reduction of the complication of these methods and processes to a collection of a few simple and clear, yet powerful, social principles. These may be arbitrarily combined and flexibly tailored to address upcoming social problems.

3.3 Orthogonalization in arts

Minimal art, a movement and style that emerged in the 1960s, stresses the idea of reducing art to a minimum number of elementary objects or primary structures, such as colours, shapes, lines and textures. In creative formal combination they produce a maximum of different pictures or sculptures. While traditional art uses an analytical approach, often attempting to represent real objects or experiences (faces and landscapes), minimal art works constructively only producing artwork that can be generated by its "construction kit".

3.4 Orthogonalization in music

Minimal music is likewise characterised by maximal simplicity and reduction of basic musical material - tones, rhythms, musical patterns, and compositorial means. The major operations to create music from the basic material are repetition of patterns, phase shifting, overlaying, stressing, adding single notes to slightly change rhythms, and sequences of tones over time. Although its underlying structure is simple, minimal music leads to a highly creative feeling of sound. It has spread worldwide in the form of techno music and its branches of trance, house, and ambient. These stress the beat and use repetitive sound patterns and rhythms to create hypnotic and ecstatic experiences.

3.5 Orthogonalization in linguistics

Umberto Eco wrote about a long-lasting search for the perfect language, a universal language in which every object, thought, idea, feeling may be unambiguously expressed (Eco 1995). A typical orthogonalization approach in the 17th century concerned a "construction kit" of forty *categories* subdivided into 251 *differences*, in turn subdivided into 2030 *species*. Each category was assigned a two-letter syllable, each difference a consonant, and each species a vowel or diphthong. So de stood for the category "element"; deb for the first element or "fire"; deba for a part of the element, a flame.

3.6 Orthogonalization in mathematics

Simple co*nstruc*tion *kits* that generate highly complex fields arise in chaos theory. Consider Julia sets that are generated by iteration of very simple functions f in the complex plane C such as

f: $C \rightarrow C$ defined by $f(x)=x^2+c$ where c is a constant.

For certain x either f^n is bounded or unbounded and the Julia set associated to f is defined to be the set of complex values x where f^n lies on the boundary.

By introducing *Kolmogorov complexity* (Li and Vitányi 1997) mathematics has formalised the notions of complexity and complication of a system. For an object or system s we define the Kolmogorov complexity C(s) to be the length (in bits) of the shortest algorithmic .description A of s. The algorithm A of length C(s) A produces s.

On the one hand, an object may be regarded simple if its Kolmogorov complexity is small, and systems are obviously complicated if their description is longer than necessary, i.e. longer than C(s). Random objects, a random sequence s of bits, say, are most complex because we are not able to find a description that is considerably shorter than the object s itself. Rather every algorithm A has to contain and enumerate each and every bit of the sequence and, thus, is about as long as the sequence itself, i.e. C(s)»|s|.

4. MINIMAL SYSTEMS IN INFORMATICS

In at least two approaches orthogonalization conquers the methodology of informatics. The first concerns the minimisation of resources: sometimes using very complicated ideas and constructions one tries to develop an algorithm that needs as little time and storage as possible to solve a problem. The problem itself may be simple in structure and easy to understand.

On the other hand the objective may be to minimise descriptions of systems. There are approaches in computer science that try to describe the most complex given structures by concepts as simple and minimal as possible or, vice versa, try to define very small orthogonal systems and then study the structures they generate. In the following we will focus on the descriptional aspects of minimalism.

We may consider informatics as a science that has developed, or adopted, the most beautiful minimal system of 0 and 1. All subjects of relevance are eventually mapped into a sequence of zeros and ones for execution by a digital computer.

There are many nice minimal systems. The following may be subdivided into two groups. One group contains minimal systems that define the executing machine or parts or models of it. The other covers systems for modelling the real world by a computer program consisting of data structures representing the static elements of the original and control structures realising its dynamic elements. For both parts of the model, depending on the underlying programming paradigm, informatics has defined fundamental minimal systems in the form of construction kits.

4.1 Orthogonalization in machines – register machines with 2 registers

The register machine (Figure 2) is an automaton whose memory contains a fixed number, say m, of registers each able to store an arbitrarily large natural number.

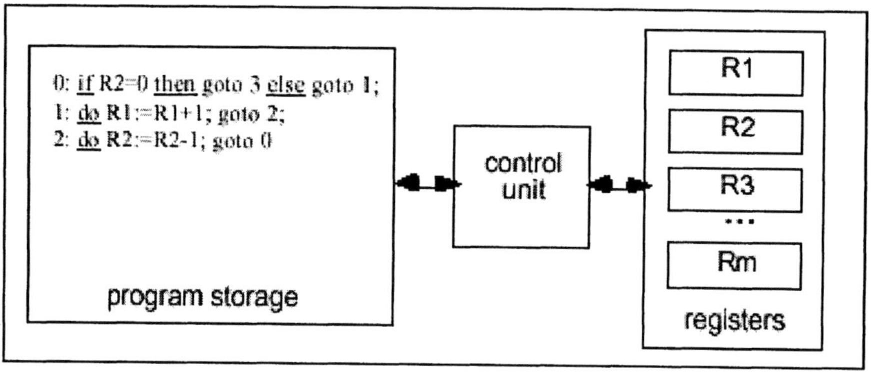

Figure 2. Register machine

The machine can perform only three operations on a register: addition of 1, subtraction of 1 if the register contains a number >1, and zero-test. A register machine program is a sequence of statements of either

i: do f; goto j
or i: if t then goto j else goto k

where i and j are labels, f is an operation on a register, and t is the zero-test on a register.

A register machine works as follows: At the beginning $r \leq m$ input values are stored in the first r registers. All other registers are set to 0. The machine executes the program beginning with statement labelled 0. The machine stops if it is to branch to an undefined label. The contents of the first $s \leq m$ registers are considered as the output. So the machine computes a function f: $IN_0^r \rightarrow IN_0^s$.

This simple machine model is extremely powerful: a register machine with only two registers can compute any computable function and so can simulate any computer. The key idea is clever coding of the storage into a single register. The coding is also an application of orthogonalization.

Orthogonalization does not stop here. A further step would be to reduce the number of instructions leading to a *single instruction computer* with equivalent power whose only instruction is

subtract 1 and jump to <label> if negative.

4.2 Orthogonalization in imperative programming – control structures

According to our observations above we can derive well-known construction kits for imperative control structures. Each of the following pairs of basic elements and operations forms a construction kit:

({assignment, goto}, {concatenation, if_then_else_fi})
or ({assignment}, {concatenation, while_do_od, if_then_else_fi})
but not ({assignment}, {concatenation, for_to_do_od, if_then_else_fi}).

The complexity of the system defined by these small construction kits may be exemplified by the following program (x is of type integer):

```
read(x);
while x>1 do
  if even(x) then x:=x/2 else x:=3*x+1 fi ;
od ;
write("I am done").
```

The program implements the Collatz function - the 3x+1- problem or Syracuse problem. For over 60 years it is not known if the program terminates for all inputs.

4.3 Orthogonalization in functional programming - λ-calculus

Functional programming is based on the λ-calculus, a mathematical formalism developed to define and use functions that are given by algorithms (Barendregt 1984). The λ-calculus is a classical minimal system that uses simple yet very powerful basic elements and operations.

Expressions or λ-terms, are defined inductively. Given a set of variables $X=\{x_1,x_2,x_3,...\}$, then it holds:

1. Each variable $x \in X$ is aλ-term (elementary λ-terms).

2. If M and N are λ-terms, then so is (MN). (MN) denotes the *application* of a λ-term M to a λ-term N, usually written M(N).

3. If $x \in X$ is a variable and M is a λ-term, then (λx.M) is a λ-term. This rule describes abstraction of a λ-term M to a function λx.M, where x is the formal parameter and M is the function body. λ stands for the keyword function in programming languages. The dot separates the function head and function body. In common with programming languages abstraction describes the parametrization of an expression like x+5 and transfer to a function f(x)=x+5. Note that a λ-term is not identified by a name and has to be completely written down wherever it is used.

Application of a λ-term to an argument is defined by the so-called β-rule which describes the replacement of formal parameters by actual ones and is defined by

$$((\lambda x.M)N)=M[x\leftarrow N],$$

where we apply the function (λx.M) with formal parameter x and body M to an actual parameter N. The rule says the entire term may be replaced by $M[x\leftarrow N]$, the term obtained by textually replacing parameter x by N everywhere in the body.

Despite its extreme simplicity λ-calculus is powerful enough to describe all computable functions, i.e. λ-calculus may be considered a fully functional programming language.

4.4 Orthogonalization in modelling – data types

The complex real world modelled by a software system usually consists of much data in many different forms and representations. Informatics has developed a method to cope with this unstructured data in a systematic way using simple construction kits. These allow any data to be described and eventually mapped into a computer.

The standard kit in programming languages is

({character, integer, real, boolean},
{aggregation, generalisation, recursion, functional spaces})

and consists of the elementary data types character, integer, real and boolean and a number of operations, so-called *constructors*, such as aggregation, generalisation, recursion, and construction of functional spaces (Figure 3).

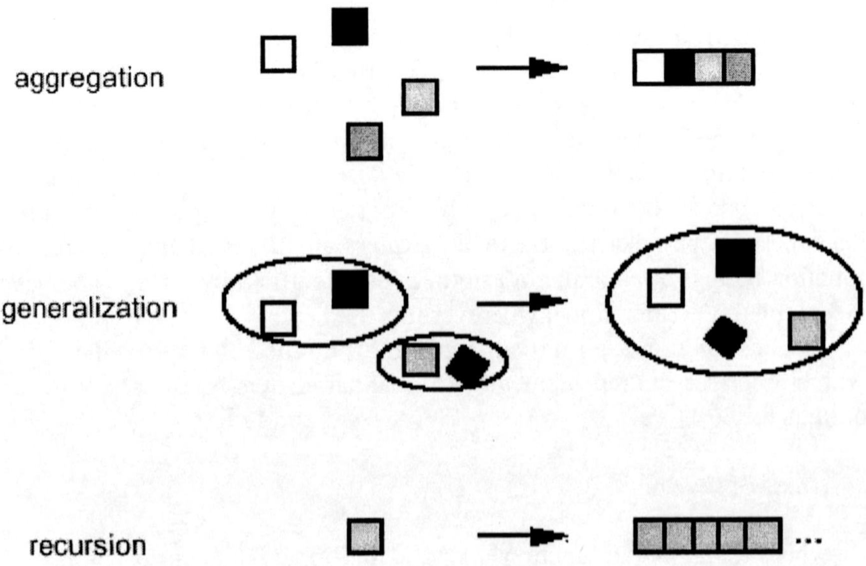

Figure 3. Effects of aggregation, generalisation and recursion

Further approaches of orthogonalization in informatics leading to minimal systems may be observed in the following:

- the universal Turing machine may be considered the one-element basis of the class of all Turing machines or even of the class of computable functions. The complexity of Turing machines or computable functions in relation to their minimal basis is best experienced by a Busy-Beaver-Turing machine;
- primitive-recursive and μ-recursive functions are defined inductively. There is a set of basic functions, e.g. constant function, successor function, and a set of operations, e.g. composition, μ-operator, that form a generating set for the class of primitive-recursive and μ-recursive functions, respectively;
- object-oriented programming attempts to develop a collection of basic reusable modules that may be configured for the concrete application area. Object-oriented design uses design patterns that form an abstract language to describe solutions to recurring object-oriented design problems;
- programming languages are often designed according to the orthogonalization principle - "When there is one good way to express something in Eiffel, there are often not two." (Meyer 1989)

- fundamental basis of the set of Boolean functions and technical foundation of computer systems are the function sets {and, or, not} or {and, not} or {nand} or {nor}. The functions of a set have to satisfy five simple criteria in order to define a generating set of all Boolean functions;
- construction kits consisting of tiny, highly modular, reusable components are being developed for operating systems. These components have well-defined operational behaviour that may be configured and functionally enriched by minimal extensions;
- homomorphisms, the bracket language and regular languages define a construction kit for the class of context-free languages such that any context-free language may be considered as a coding of correct bracket expressions with some regular languages in between;
- to prove that a system is not orthogonal one often uses the idea of *emulation*: Given a generating system, if one of the basic elements or operations may be realized by the others then the system is not minimal.

5. MINIMAL SYSTEMS IN INFORMATICS LESSONS

The above examples support the thesis that orthogonalization is a fundamental idea of informatics that satisfies both the Horizontal Criterion and the Criterion of Sense. If so the idea must be integrated into informatics lessons and taught in a way that clearly illustrates its property of linking different informatics subjects and integrating diverse phenomena and methods under a common concept.

To achieve this orthogonalization and minimal systems should be dealt with in schools according to the following guidelines:

- Orthogonalization should be introduced at every stage of intellectual development and, following the spiral principle, examined with increasing level of elaboration and formalization - starting in primary school.
- It must be made clear where exactly the idea should be introduced and applied in the informatics curriculum and what advantage its use offers for solving the problem.
- The appropriate minimal system, its basic elements and operations, and their inter-relationships must be explained in detail.
- The inner structure of the field defined by the minimal system must be analysed.

- It is important that students are shown how this approach and earlier applications of minimal systems in other subjects coincide or differ.

The key concept is the Vertical Criterion. That enables the idea to be taught on every school level "in some intellectually honest form". Some examples mentioned above may be reduced to the primary school level.

From the methodical point of view orthogonalization appears to be a valuable subject. Students usually have gained early experiences with construction kits in their everyday life e.g. LEGO. They may easily grasp small systems with few orthogonal basic elements and operations and, via constructivism, experiment with basis and operations and explore, possibly in project work, the space "spanned" by the construction kit and experience its vast complexity.

REFERENCES

Barendregt, H. P. (1984) *The Lambda calculus*. North-Holland.

Biedenkopf, K. (1994) Komplexität und Kompliziertheit. *Informatik Spektrum* 17, pp. 82-86.

Bruner, J. S. (1960) *The process of education*. Cambridge, Mass.

Eco, U. (1995) *The search for the perfect language*. Blackwell.

Gamma, E., Helm, R., Johnson, R. and Vlissides, J. (1995) *Design patterns*. Addison Wesley.

Li, M. and Vitányi, P. (1997) *An introduction to Kolmogorov complexity and its applications*. Springer.

Meyer, B. (1989) From structured programming to object-oriented design: the road to Eiffel. *Structured Programming* 1, pp.19-39.

Miller, G. A. (1956) The magical number seven plus or minus two: Some limits on our capacity for processing information. *Psychological Review* 63, pp. 81-97.

Schwill, A. (1993) Fundamentale Ideen der Informatik. *Zentralblatt für Didaktik der Mathematik* 1, pp. 20-31.

Schwill, A. (1997) Computer science education based on fundamental ideas. In *Information Technology - Supporting change through teacher education*, D. Passey and B. Samways (eds.), Chapman Hall. pp. 285-291.

ICT in Education: Aspirations and Tensions

David Wood
Institute of Learning Sciences, University of Nottingham, Nottingham NG7 2RD, United Kingdom

djw@psychology.nottingham.ac.uk

Key words: Education, Future, ICT, Learning Models, Policy Makers, Recommendations

Abstract: Eight 'axes of tension', around which developments in the exploitation of ICT will revolve to achieve aspirations for schooling, are identified and explored in this paper. Possible interactions between these tensions and aspirations with the educational promises held out by the technology and constraints on change and innovation are illustrated with three main scenarios to portray possible futures for schooling. The main aims of the paper are to motivate and direct debate, and to provide access to a framework that others may use to etch out and explore their own preferred visions for the future of education.

1. INTRODUCTION: THE THINK PROJECT

Initiated by the European Schoolnet (EUN) and sponsored by the European Commission, the THINK project was designed to articulate and explore policy options in six national contexts, and an EU-wide one, for the exploitation of ICT to develop and strengthen school systems (Wood, 2002). The focus was on future educational *uses* of ICT, not provision of technical infrastructure. The time scale to be envisaged was short - three to five years.

Four scenarios were identified:

In **scenario 1**, ICT serves to augment the centralised control of all aspects of schooling, strengthening state regulation of curricula goals, content, delivery, pedagogy, formative and summative assessment, accreditation and finance.

In **scenario 2**, policy makers acknowledge the fundamental uncertainty concerning the nature of the impact of ICT on learning and teaching processes. Any state controls are relaxed to engender and support a research and development role for teachers and schools. Schools strive to become knowledge producers and learning organisations.

In **scenario 3**, the most radical future, the role of schools as organisations designed to promote learning how to learn, responsible citizenship and as key nodes in new communities is brought to the fore as their function for knowledge transmission is downplayed.

A **4th scenario**, in which ICT fails to deliver and 'melts down', considers various possible failures in policy (in)actions that undermine attempts to innovate with ICT.

The first three scenarios overlap with those etched out over a much longer time horizon by OECD (2000; 2001). The THINK project thus provides a means of mapping aspects of policy options for the short-term future onto potential longer-term changes in school systems.

2. AXES OF TENSION

When ICT becomes mission critical for educational systems, several processes will be set in chain that will generate new tensions and dilemmas for schools and education authorities. The main dynamics driving innovation in schooling stem from strategies adopted to resolve a set of "axes of tension" created by the impact of the technology on learning.

2.1 Axis 1 - Innovation with ICT is inhibited and stifled by failure to re-think the curriculum.

Propositions: The definition of what is worth knowing and the skills and tools implicated in what it is to be knowledgeable are being transformed by the impact of technology on all aspects of professional, vocational, private and public life. There is a growing tension, and an inherent contradiction, between demands for radical change in educational priorities and processes and the expectations and goals embodied in school curricula.

The current trend is simply to pile new learning objectives and teaching responsibilities on top of the traditional ones. This creates an unnecessarily heavy load, undue stress and a confusion of aims for schools.

In scenario 1, this tension is refuted. The belief is that the curriculum can remain relatively intact, although there is some extension of the educational pathways available in the later years of schooling. The view taken is that through appropriate changes in pedagogical practice schools can achieve both the old and the new priorities.

In scenario 2, the tension is accepted. A strategy is developed that preserves a core of traditional curricula goals and assessments but also creates significant space for innovation. School achievement is assessed by two independent means. One is based on national examination and the second on innovative, school-based assessments. Schools have the task of innovating in order to meet the new educational priorities.

In scenario 3, the traditional conception of school learning as primarily concerned with the mastery of a pre-determined body of knowledge is rejected. The overarching objective is to create and develop a system in which schools become the core centres of community life and the development of skills in self-managed learning, critical thinking and responsible citizenship form the major goals of schooling. Decision-making about specific educational goals, strategies and practices are primarily the responsibility of the school and its community. Their task is to translate overall priorities for education into practice. ICT networking is exploited to engage the expertise of different communities in new ways of assessing the value and impact of innovative approaches to learning and teaching.

2.2 Axis 2 - Innovation with ICT fails, being inhibited and stifled by a failure to capture the imagination and support of parents and the public.

Propositions: A tension exists between the need for radical change demanded by educational policy and the majority sentiments of parents, the media and the electorate. Attempts at educational reform are currently ignoring this tension, remaining too inward looking and failing to exploit the technology to modernise the policy making process itself.

In scenario 1, this tension is not acute in the short term since the curriculum and examinations reflect the status quo and have majority support. Investment in ICT and in teacher training rests on the promise that it will enable the extended educational priorities to be realised through improved teaching and learning. ICT networking is used widely in attempts to maintain the support and educational involvement of parents.

In scenario 2, public confidence will be supported by the fact that recognisable elements of the traditional curriculum and certification are maintained in the early stages of reform. Strategies to capture the public imagination and support include the use of networking to communicate and disseminate innovative uses of ICT for teaching and learning identified in the course of peer assessment by teachers.

In scenario 3, the use of ICT for extensive, deep and sustained networking between schools and the community are central to this educational system, so the nature and rate of innovation is directly influenced by parental and public support. One role for the educational authorities is to exploit ICT to help in identifying, celebrating and disseminating innovative examples of the new schools.

2.3 Axis 3 - Innovation fails because the tools and practices used to monitor and guide learner progress and achievement do not support the achievement of new objectives.

Propositions: The tools and practices used to assess learning and teaching are concerned with the memorisation, reproduction and mastery of approved bodies of knowledge. A failure to innovate and change these tools and practices inhibits the system's ability to realise innovative uses of ICT. Teacher training, and demands for change in pedagogical practice, are undermined by the fact that most teachers will appeal to the need for learners to succeed on traditional forms of assessment as a reason not to change. It will prove impossible to guide and coach the learning process effectively without new means of formative assessment. This will create confusion and resistance from learners, teachers and parents.

In scenario 1, this tension is not recognised initially. Later, regimes of national testing and the use of on-line assessment will be developed to acknowledge and support the development and assessment of new skills.

In scenario 2, since teachers and schools will become centres for educational research and development, the task of discovering and perfecting new approaches to learner guidance and formative assessment will rest on them. These will be validated against school-based assessments of learner achievement.

In scenario 3, this tension is not recognised since schools and the community will be tasked with the job of developing their own methods of assessment and guidance.

2.4 Axis 4 - The innovative use of ICT will increase the gap between high and low school achievement

Propositions: ICT is used to open up more choices for the learner. This leads to an increase in the magnitude of the gap between high and low school achievement, leading to accusations that innovation in schools is promoting greater inequalities in educational opportunity and life chances, contributing to social exclusion and weakening social cohesion.

In scenario 1, the system seeks to resolve tensions by making schools accountable for minimal levels of educational achievement in key areas. It exploits ICT to strengthen and defend regimens of selection to accommodate the increasing variability in rates and directions of pupil progress.

In scenario 2, greater emphasis is placed on identifying and supporting individual potential. The aim is to minimise apparent under-achievement by expanding the range of opportunities to succeed and by enhancing the system's ability to offer sound guidance and advice. This leads to a greater diversity of provision, more varied educational pathways and a wider definition of educational success.

In scenario 3, individual differences in achievement do *not* figure strongly in educational accounting. The approach to assessment adopted is itself innovative. More emphasis is laid on the value and evaluation of contributions to the community and the development of responsible citizenship. Assessment is based on the documentation of individual and group portfolios and includes evaluations from school and non-school sources.

2.5 Axis 5 - Teachers as technicians versus teachers as professionals.

Propositions: Teachers are in the front line of attack of attempts to implement policies designed to reap educational benefits from investments in ICT. Innovations in the effectiveness of teacher recruitment, training and professional development are key issues and will remain so.

In scenario 1, the State defines in detail what is worth knowing, and prescribes how the assessment of knowledge and skill will take place. The role of teachers is to optimise their practices in order to maximise the fit between the stated goals and the measured outcomes of schooling. Teachers are essentially technicians.

In scenario 2, the professional roles of teachers expand to become far more demanding. They are tasked both with maintaining standards on nationally defined assessments of pupil achievement and with a research and development brief. Discovering how ICT can be exploited to realise the enlarged set of goals for education prescribed by the State is largely their responsibility. They must discover and demonstrate how changes in pedagogical practices and innovative forms of assessment can support the achievement of each learner's potential.

In scenario 3, the extent of the teacher's responsibility is even greater. Individual schools and teachers are professionally accountable to learners, parents and the community for defining the content of the curriculum and for the documentation of evidence for what pupils learn and achieve. Additionally, teachers must work with their school to develop and sustain close community links as they collaborate with others to identify and realise the locally defined objectives for schooling.

2.6 Axis 6 - Learners as consumers versus learners as an asset.

This tension invites the most radical re-thinking about the role of schooling.

In scenario 1, learners and the taxpayer are viewed as consumers of educational products and services. The main role of schools, as providers, is to equip learners with the knowledge, skills and attitudes they will need to function as productive and prosperous future citizens thus minimising social exclusion and contributing to social cohesion. Taxpayers' contributions are defended as an investment in that future. Schools are held accountable for giving value for money by achieving standards that, in the State's view, are held to be the best and most reliable indexes of future potential.

In scenario 2, with its emphasis on schools as centres for creating and disseminating knowledge about learning and teaching, a similar view of learners and families as consumers prevails, albeit within the context of a broader and more differentiated definition of educational achievement.

In scenario 3, the major emphasis lies on learners as an asset to the local and wider communities. With a greatly reduced burden of knowledge transmission, learners and schools become "time rich". Learners are viewed and treated as junior citizens with the time, motivation and potential to make significant contributions to the social, cultural, intellectual and economic lives of their communities. Contributions are co-ordinated with the pedagogical goals of helping learners become more independent and better able to collaborate in identifying and tackling complex and authentic tasks. In so doing they will become critical, informed participants in national life.

2.7 Axis 7 - The burden of maintaining high quality of ICT provision exceeds available public funding.

Propositions: The cost of maintaining and upgrading ICT infrastructure and services in order to build sufficient capacity and redundancy into technological and human resource provision for schools is currently under-estimated. A realistic assessment of the levels of recurrent investment required will only become clear over the next five years when the technology becomes mission critical for schools.

In scenario 1, the educational system is best placed to augment public funding with private financing. Given the widespread use of common content in the delivery of the national curriculum, households present a large and recurrent market for ICT products and services. This supports a buoyant domestic industry that is then able to provide for the less lucrative school market. Teacher training provision as an element of commercial service provision becomes common freeing the State from the costs of updating and re-training teachers. The State is able to target additional resources for enhanced support to less well off schools and households.

In scenario 2, maintenance of the infrastructure remains a State responsibility but there is the expectation that, within five years, teachers will become the major producers of content and services to support educational uses of ICT. Expertise grows within the educational system leading to less reliance on external provision and support and to costs lower than those faced in scenario 1. ICT supports the continuous professional development of teachers reducing training costs as this function is served by professional networking.

Scenario 3 is the most demanding on public approval and financial support. Schools will rely heavily on gaining 'free' support and expertise from local communities. Where significant variation exists in the levels of prosperity of different communities these will be reflected in the levels and quality of ICT available to schools. The State will attempt to compensate for any such differences through selective regional funding.

2.8 Axis 8 - Protection versus Censorship.

Propositions: ICT creates enduring tensions between the goals of providing learners with access to the information society and the responsibility for regulating their own learning on the one hand, and the need to protect them from potential exploitation, indoctrination and crime on the other. Open access by young learners, and the potential for undesirables to have open access to them, raises the potential for morally repulsive, politically subversive and economically dishonest and illegal contacts. As never before, it is possible for the young to be exposed to the more mature in unaccompanied and unguided non-public spaces. Developments in the technological capacity to track and model the journeys that learners make as they explore and learn also leaves them open to new forms of exploitation.

In scenario 1, the potential for both state protection and censorship are strongest. The state exerts direct control over the content and functions of ICT in schools and attempts to become the gatekeeper to the world of information. There will be a need for specific regulations or guidelines concerning the ownership of the information and intelligence gained from a learner's use of ICT and the rights of access to such information.

In scenario 2, the responsibility for placing constraints on access and privacy rests with schools and parents, albeit within a regulatory State framework. Since ICT content and function will develop within the schools issues such as intellectual ownership, copyright violation and access to personal information will be taxing issues for local management.

In scenario 3, the moral, legal and economic dimensions of ICT will be an integral part of the intellectual concerns of learners, schools and the local community. It is likely that early but guided and critical exposure to potentially harmful materials will be a feature of schooling from the early years. The timing and nature of such contacts will be determined at a local level in collaboration with parents and the community within a national framework of regulation. It seems likely that such an approach will only be tenable within a highly coherent system.

3. CONCLUDING COMMENTS

The axes of tension and scenarios developed in THINK represent imagined, possible futures generated by the interplay between educational aspirations, the promised powers of ICT to promote change, and sets of levers and constraints on change and innovation. Although the six areas of aspiration and the seventeen sources of constraint identified in THINK may be culture specific, the claim is that the eight axes of tension just outlined will challenge any education system bent on change.

The scenarios alluded to were not formulated as descriptions of likely futures, nor as predictions, nor yet as prescriptions for action. The aspiration is that the 'analytical tool kit' generated by THINK will motivate and support others as they try to envisage and bring about futures for schooling that they, and their society, come to view as desirable.

REFERENCES

OECD in collaboration with CERI and the Netherlands Ministry of Education, Culture and Science (2000), Schooling for Tomorrow. An OECD/Netherlands Conference. Rotterdam, November.

OECD (2001) *Education Policy Analysis 2001*, Centre for Educational Research and Innovation.

Wood, D. (2002) *The THINK Report: Technology in Education Futures for Policy.* [http://eminent.eun.org/THINK_FULL_DRAFT_2pp.doc]

e-Learning Technology: Convergence with the Mainstream

Colin Harrison
IBM Institute for Advanced Learning, IBM Zürich Research Laboratory, 8803, Rüschlikon, Switzerland

ch@zurich.ibm.com

Key words: Learning Objects, Sequencing, Interaction Modelling, Web Services

Abstract: The evolution of technology-based learning systems over the last thirty years has been driven by the desire to implement increasingly rich pedagogical models. This has led to remarkable innovations in interaction, collaboration, and the use of rich-media in learning systems. However, these technology innovations have rarely been absorbed into the mainstream of information technology and consequently have failed to evolve into broadly implemented standards. Abstractly we can view learning as a sequence of encounters among instructors, learners, and information. These encounters may be synchronous or asynchronous, local or distant, formal or informal, solitary or en masse. Computer Science is seeking to develop technologies that will support a new set of Web experiences. These technologies will provide a generalised model for describing encounters among people and information and are based on open industry standards. The application space for these experiential technologies will eventually include e-commerce, professional communications, games and entertainment, but learning appears likely to be the first practical application.

1. INTRODUCTION

For over 20 years we have been installing personal computers in classrooms for primary, secondary, and tertiary education in the deeply rooted belief that this will improve education. While the computer is clearly an important tool for 21C knowledge workers, this has been true of many 20C inventions that have not been so strongly embraced. Moreover, today at least, many learners have better access to computers at home than at school or university. There is something intensely compelling about the personal computer that convinces many people that it must *a priori* be a valuable tool for learning.

Some speak of technology-based learning - hereafter e-Learning - as the manifest destiny for the transformation of our educational systems and then seek to address "obstacles" to this transformation. This perspective ignores the distress in educational systems and prompts the question: If e-Learning is such a powerful technique, what critical problems in training and education can it solve? Practitioners lack professional development opportunities to get beyond the practicalities of how to drive a Web browser and into a deeper understanding of how this technology will inevitably transform educational methods. These two communities need much deeper engagement to take our instinctive belief in the educational value of personal computers to a level where we can talk factually about the relative merits of investment in e-Learning versus investment in other education facilities.

In the work of IBM's Institute for Advanced Learning, we view the computer as an active partner in a dialogue that includes instructor, students, mentors, and other parties. We view the learning process as a collection of encounters between the individual student and the instructor, the body of knowledge, other students, and a variety of learning methods including discovery, inquiry and apprenticeship. The instructor orchestrates these encounters to enable each student to reach the desired learning outcome. We view the computer as a facilitator for many of these encounters, adding to the traditional classroom experience the benefits of Information and Communication Technologies (ICT). The instructor should orchestrate this facilitation as part of the overall learning process.

Our key goals are:

- to provide leadership to educators and industry in developing a set of technologies and content based on open industry standards to fully develop the capabilities of e-Learning;
- to work towards a factual basis for the application of these capabilities to produce specific learning outcomes.

The result should be that educational practitioners extend their learning methods in pedagogically sound ways supported by tools and content that are meaningful in the learning context.

These are extremely large and complex problems so we seek key leverage points where IBM is well placed to produce movement towards these goals.

2. HOW CAN COMPUTERS HELP LEARNING?

Let's start with a simple question. In what ways do we think computers can be helpful and add value in learning? This has been widely debated and the answers suggest it has something to do with:

- access to information, both local and also beyond the boundaries of the learning institution;
- structuring frameworks for interactions among instructors, learners, and tutors;
- enabling communication among a group of people who cannot be at the same place at the same time;
- interaction with simulations and games, enabling new forms of learning experiences;
- individualised learning;
- new methods of learning & assessment;
- providing communities of interest for instructors, parents, and learners.

Note that we are not looking at the computer as a vehicle for pedagogy. This role falls to the instructor, who may employ the computer to support various activities within a learning method.

How does the computer contribute to these activities - what high-level capabilities of the computer support these? Here is a partial list:

- Web browsing and browsing other documents, whether local or remote, the ability to access and experience digital information.
- Multimedia, the ability to deliver digital information in the form of moving pictures and sounds.
- Interaction, the ability of the computer to modify its actions in response to input from the user. This capability is basic to all forms of learning experiences. A specific form of interaction is the ability of the computer to modify its actions based on a simulated interaction with a virtual environment, such as a flight simulator, or with a human partner, such as an injured patient. Closely related to these interactions are games, where players work (competitively or collaboratively) towards a goal.
- Adaptation, the ability of the computer to adapt its interaction with the user to optimise the experience it provides - in terms of learning or completing a transaction - to the user's needs.
- Collaboration, the ability of a group of people to have common views of documents, multimedia content, and chat. Examples include one-on-one dialogues for learning and coaching, collaborative problem solving, and guided use of a computer application.

- Authoring, the ability of the computer to manipulate multimedia and other content at the direction of a user, typically an instructor, to produce a meaningful learning object. At its most basic this is the capture of primitive content – text, image, objectives or state. At its most advanced it may begin to provide pedagogical support for specific learning goals.
- Content management, the ability of the computer to act as a library for digital content of all forms.
- Testing, the ability of the computer to pose challenges to an individual learner (or group of learners) and determine whether their response indicates understanding of the subject matter.
- Attention, the ability of the computer to assess the mental state of the user and to communicate this to a person, typically an instructor or tutor.

The last item is the weakest link. One of the greatest losses in adopting electronic methods for learning is the dramatic loss of feedback among the several parties on their mental and emotional states. In face-to-face interactions as humans we express these states through many visual signals. There are many attempts to compensate for this loss, for example through the use of bi-directional video or through animated avatars. This is a significant problem for e-Learning as skilled instructors make great use of the verbal and non-verbal feedback they receive from a face-to-face audience in adapting, repeating or restating information. The loss of that feedback mechanism is one of the great challenges that instructors using e-Learning face.

The first four capabilities are closely related to activities that have been demonstrated over many years to be key activities in enabling the learner to capture information and turn it into knowledge. What is new is that now the computer supports these activities. In recent years, e-learning practitioners, especially in corporate training, have recognised the need to apply all these various techniques in "blended learning".

Figure 1 illustrates blended learning. It chunks learning activities into four main layers - learning from information, learning from interaction, learning from collaboration, and learning from co-location (face-to-face). This provides a useful framework in which to place a variety of computer-supported learning methods - it does not express a particular learning model. As we move vertically, we go from learner-led, individual-study, based on electronic media up to instructor-led, multiparty, face-to-face learning.

The left-hand side of Figure 1 focuses on "formal" learning; learning activities guided by an instructor towards some goal. Studies have recognised that learning does not stop with completion of the course, but continues through informal learning experiences. These ultimately deliver the majority of our working knowledge. We are concerned to provide computer support for these informal learning activities illustrated in the right-hand side of Figure 1. Moving horizontally, we go from formal, structured learning, often long-term or "Just-In-Case" learning usually performed in an office setting, to informal, unstructured, often short-term or "Just-In-Time" learning that may take place at any time or anywhere. Figure 1 summarises the range of formal and informal learning activities – independent of pedagogy – for which we seek computer support. From this we can advance to asking what technologies are required to provide this support.

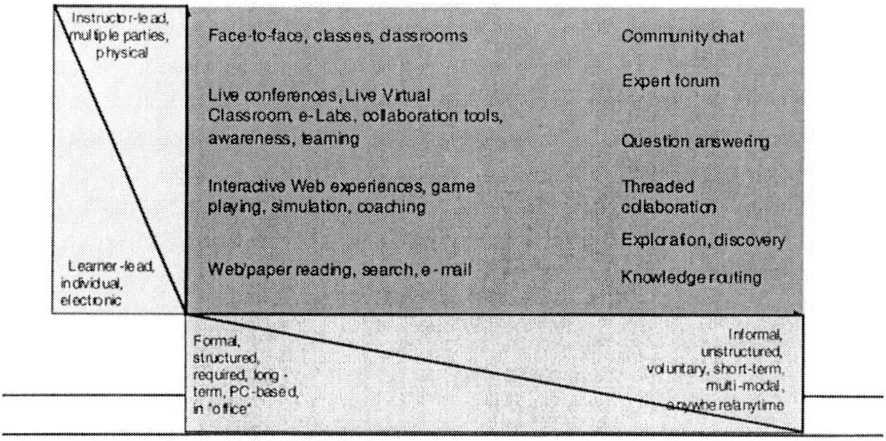

Figure 1. The blended learning model

Given this repertoire of capabilities can we develop prescriptive methods for combining these capabilities to compose learning activities that support a variety of teaching models? Such activities wrap around into the uppermost levels of the technology.

One last dimension is how the e-learning activity is delivered to the end-user. While desktop or laptop computers are the principal client devices, we anticipate that television will evolve into an interactive medium, capable of providing many, if not all capabilities discussed above. Networked game platforms (Sony PlayStation © and Microsoft X-Box ©) could, in principle, support these capabilities. Lastly, mobile devices such as PDAs, wearable computers, and 3G telephones may support sub-sets of these capabilities that have very high value for specific learning needs.

3. TECHNICAL FOUNDATIONS FOR E-LEARNING

e-Learning systems exemplify distributed computing. Compared to many other forms of computing e-Learning has two special emphases - on content rather than transactions and on human-human interactions rather than purely human-machine interactions. Computer-based training pioneered many innovations in interactive systems, multi-media, and virtual reality. In many cases it has been too early in its adoption of these technologies and they have had to be implemented in the e-Learning applications, rather than being used to exploit facilities provided by the operating system or middleware. Consequently, earlier e-Learning systems have appeared to be a distinct class of computing, requiring its own unique technologies and standards.

However, e-Learning technology will soon converge with other content-centric or interaction-centric activities to become "just" an instance of multi-party, network-mediated, adaptive, interactive, rich-media applications. Broadband distribution, digital media, XML composite "documents", collaboration, unstructured information management and digital rights management are all evolving to support these complex applications that e-Learning needs. In response to broad-based needs, these technologies are on track to provide the primitives needed to implement e-Learning applications.

A key activity is ensuring that the requirements of e-Learning are incorporated in the relevant standards and implemented in products. We must partition these needs into the generic (requiring configuration to the activities in question) and the specific (to the activity). The e-Learning community must converge with these mainstreams and use commoditised technology. e-Learning will no longer be a distinct class of computing.

The driving force is the emergence of other multiparty, network-mediated, adaptive, interactive, rich-media applications - e-commerce systems that adapt their behaviour to the customer's preferences, negotiating or bidding systems that structure the dialogue among the parties, and games, a structured set of interactions among players. All of these specialise, but a key area of technology work lies in abstracting the generic mechanisms required to model these and other human-human and human-machine interactions. The emergence of these several forms of multiparty, network-mediated, adaptive, interactive, rich-media applications leads to the emergence of mainstream technologies that, in particular, are capable of supporting e-Learning. A clear, though unfortunate, example of this is the enormous overlap between metadata specifications for e-Learning Objects on the one hand and for interactive rich-media on the other (Learning Resources 2002; IEEE 2002; MPEG-7 2002). I refer to this as "convergence".

Today we are close to a good set of distributed computing primitives based around open industry standards. We have ideas about how to assemble these into multi-party, network-mediated, adaptive, interactive, rich-media primitives for various purposes, including e-Learning (Figure 2). The foundations comprise the distributed computing primitives that are generally accepted as common elements for almost all the mainstream applications. However we should be trying to push convergence as high up the stack as possible. We do not need a multiparty collaboration primitive specialised on e-Learning but we need a generic multiparty collaboration primitive that could be configured to support various learning interactions.

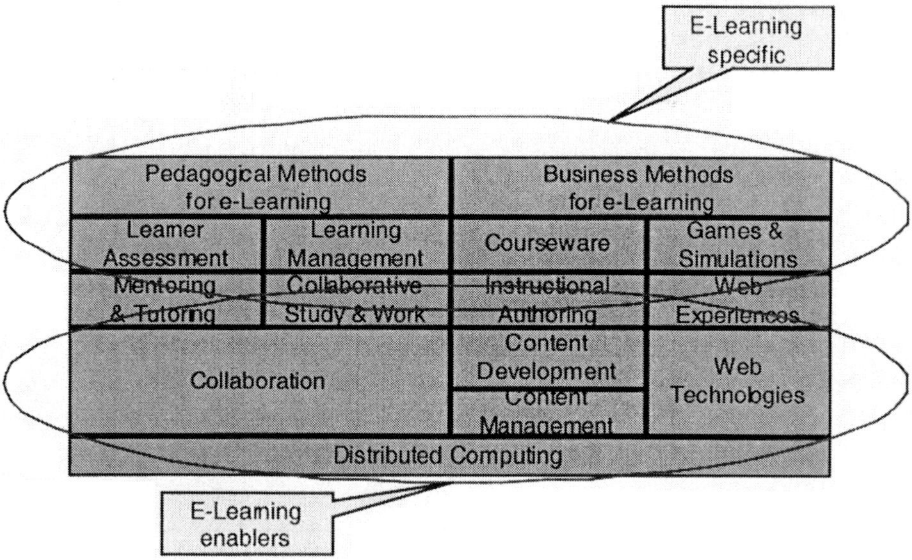

Figure 2. Building blocks for e-Learning

Above the Distributed Computing layer lies convergent middleware that provides support for generic multiparty, network-mediated, adaptive, interactive, rich-media applications. This convergent middleware includes collaboration, content development, content management, and Web technologies. Above this layer are e-Learning-specific applications that leverage other primitives to compose complex applications that support pedagogically sound learning activities.

We require research programs that bring together Computer Science (CS) and Human Computer Interface (HCI) people working bottom up to study how to generalise the application of the primitives into bigger chunks that can be configured to represent, say, e-Learning, e-Commerce, e-Negotiation or e-Games.

If we take a top down view, there is work in which we ask practitioners - instructors, tutors and mentors - to engage with the CS/HCI community on how to most effectively apply these chunks to model electronically their best behavioural practices. That is, models of interaction, spanning the two dimensions of blended learning shown in Figure 1, and supporting a variety of pedagogical principles. These models act as maps not rigid processes illustrating ways the learner may reach the learning goal. The learner, guided additionally by an instructor, a coach, or peers may explore and follow any route. Each domain: learning, commerce or negotiation has half a century of behavioural research behind it that is collected into a finite set of models/principles - discovery-based learning, inquiry-based learning and cognitive apprenticeship.

I do not believe that it is the job of the e-Learning technology community to figure which model is best: that role belongs to the instructors. Our job is to figure out how the chunked primitives should be configured to support each of these models.

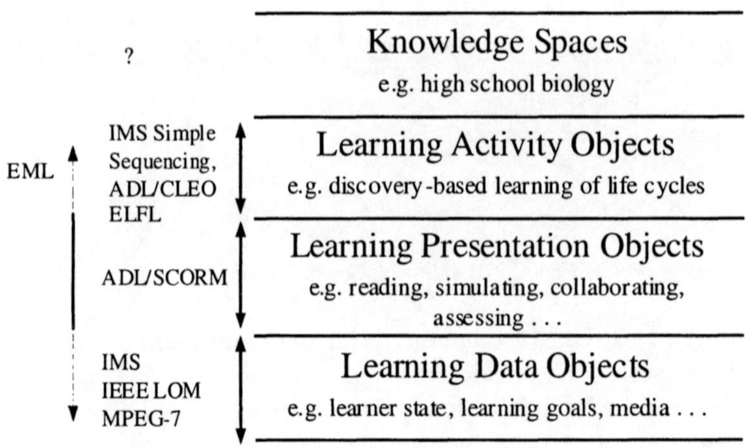

Figure 3. e-Learning structures showing the hierarchy of learning objects

We must judge what is most effective, since inevitably when we transfer these models to electronic interaction they change subtly. We must work with the practitioners to determine how to improve fidelity to the original models or exploit the new characteristics. Let us first develop the capability - building on the work of the Educational Modelling Language (EML) group at the Open University of the Netherlands.

In addition to functional support for the delivery of learning activities, we also need technology support for structuring of digital content that creates these experiences. Figure 3 shows a hierarchy of such structures, commonly referred to as learning objects. At the base are the data primitives of e-Learning: rich-media content, learner state and learning objectives (Learning Data Objects in our terminology). In the second layer these are composed into short learning experiences (units of study in EML terminology - Learning Presentation Objects in ours). The third layer is the composition of learning experiences into extended learning experiences, lasting for an entire session, class, or course. In the topmost level there is the structuring of knowledge itself. Figure 3 also shows the coverage of various learning object standards over these layers.

Structuring Learning Data Objects is the basic digital library problem: annotating digital content with metadata that allows it to be stored and retrieved. The IEEE Learning Object Metadata (LOM) standard has recently been adopted for this purpose (IEEE 2002). Structuring Learning Data Objects into Learning Presentation Objects is addressed by both the EML (Koper 2001) and Shareable Content Object Reference Model (SCORM) (ADL 2002) specifications from the Advanced Distributed Learning Network (ADL). The IEEE LOM standard and the ADL SCORM specification are widely adopted in commercial learning systems.

The structuring of learning experiences (Learning Presentation Objects or EML Units of Study) into extended learning activities (Learning Activity Objects) is an active area of exploration. This is being approached in terms of sequencing Learning Presentation Objects into graphs whose nodes are the learning experiences. The exit from each node - at the completion of the learning experience - is connected to one or more subsequent nodes. The selection of which subsequent node is visited may be made to depend on in-line assessment or other learner-centric factors.

One element missing from these specifications and standards is a strong architectural model. In our work we consider both the structuring of multiparty, network-mediated, adaptive, interactive, rich-media learning activities and the end-to-end delivery platform as well as the content lifecycle. Our project on the sequencing problem is known as the e-Learning Flow Language (ELFL) (Farrell 2002). It adds to the existing Learning Data Objects and Learning Presentation Objects the concept of Learning Control Objects. Their function is to enable the control of the flow of learning experience across the graph of Learning Presentation Objects. These varying paths may be made dependent on such factors as the learner's prior knowledge, the learner's role (practitioner, manager or inspector), or the learning need ("Just-In-Case" or "Just-In-Time"). Each node or learning experience is modelled as a Web Service, which connects us to a rich set of distributed computing primitives, such as the Web Services Description Language (Web Services 2002).

We have reviewed this direction of research with IMS, ADL and academic researchers. It is our intent - when we have working code - to introduce these concepts into the open industry standards processes.

Above these models of interaction there is the background or context for the learning activity. A key goal of deep learning is not merely the memorisation of information, but the generation of the mental framework or pigeonholes of a particular discipline into which new information can be rapidly assimilated. This is often referred to as a "mental map" - from the point of view of a person - or "knowledge map" from the more abstract view of the discipline itself. One traditional tool for representing the knowledge map is a book, which presents the information on say, physics, in a logical sequence that can be comprehended by the learner. Another common knowledge map representation is the organisation of books in a library.

A syllabus can be represented as a taxonomy or a graph. This may be a good mechanism for structuring a collection of learning activities that provide syllabus coverage. A graph could include constraints in the form of prerequisites to launching a given activity but a graphical representation of this kind may not be the most engaging or memorable way of conveying the structure of the knowledge space.

What kind of computer-supported modelling can we apply to help the learner develop the mental map? Such work invariably involves various ways of visualising the relationships among information elements of a discipline like ontological representations of concepts, virtual reality spaces, and adaptations of real spaces (Sherman 2002; Cetkovic et al 2002).

Lastly we need a set of authoring tools that enable practitioners to manipulate these electronic models in terms of the models of their own profession. These tools must provide support for the practitioner at four levels:

- finding the learning data - the basic information;
- composing this into a meaningful and engaging learning activity according to some pedagogical principles;
- sequencing learning experiences into an overall set of learning activities to cover a syllabus;
- integrating several learning activities into a visualisation of the knowledge space.

This is a very difficult problem. Today we can do a poor-to-fair job on finding the learning data and we have template-based tools for composing these into learning experiences. Better support for pedagogy in creating learning experiences is a frequent customer request. We have little or no idea of how to use the emerging sequencing techniques. Despite experimentation on visualising knowledge maps there is no codification of how to effectively apply these. If this seems hard from the perspective of technology support, it appears incomprehensible from the perspective of most practitioners. No wonder they are confused about learning objects. Practitioners and technologists together need to do a lot of exploration and codification of how to use these new structuring methods. Then we have to make these available in terms of concepts that are meaningful to the practitioners through tools that are as easy to learn to use badly as Microsoft Powerpoint ©.

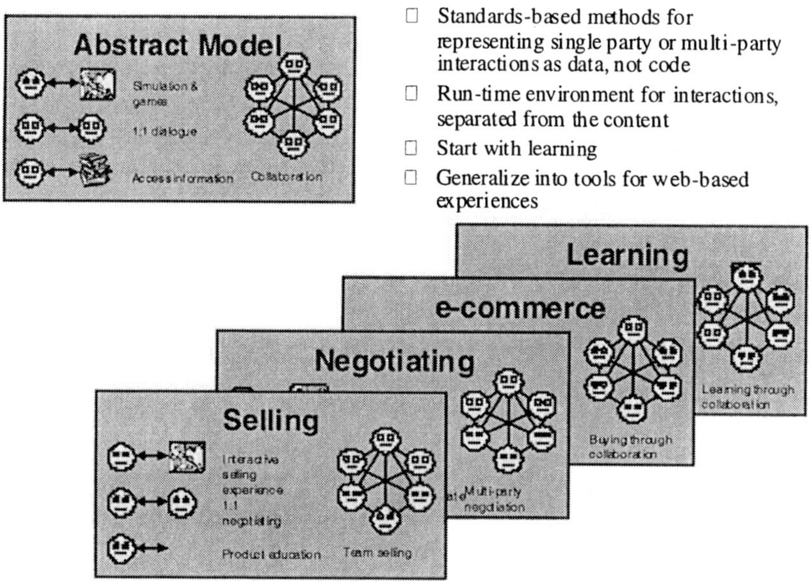

Figure 4. Modelling human activities

Fortunately we are not alone. The model we have for the application of computers in support of learning is that of a collection of mechanisms for enabling interaction among one or more learners, an instructor, a body of information, a group of peers, and possibly a coach. This is a very finite set of activities. I earlier indicated a number of fronts where Computer Science is intersecting with Behavioural Science with the aim of modelling human behaviour. In e-commerce we seek to make Web sites that are less frustrating to use because their behaviour conforms to the user's mental task model. In negotiation we model the progressive disclosure of information between, say, a safe driver needing insurance and one or more insurance providers. As Figure 4 shows, in the next years we expect to develop generalised methods for supporting these kinds of behavioural modelling which can then be specialised to an area such as e-Learning. This is important. It finally connects e-Learning techniques and technologies to the mainstream of commercial software development. This will attract the benefit of a much larger investment stream.

We believe the key areas for technologies for delivering the e-Learning experience in the short term are:

- Ensuring that mainstream, converged tools and middleware support the generic requirements of e-Learning as converged primitives in areas such as content management, collaboration and authoring.
- Separating out the purely e-Learning concerns and developing ways to express these through lower-level, converged primitives. We see ELFL as an example of this – leveraging Web Services primitives to create educational experiences.
- Bridging the semantic gap between low-level content manipulation and high-level lesson/course design. The authoring burden on e-Learning practitioners is extremely high. To fulfil even the more modest visions, we must find ways to enable instructors to focus on the problems of learning and to use e-Learning tools in that context.

4. SUMMARY

Our view of the role of computers in learning is that they provide a repertoire of capabilities from interacting with information through simulated experiences to collaborative problem solving and coaching. These capabilities are rooted in the mainstream technologies of distributed computing which are being used to support relatively new forms of computing applications that are intensive in the use of rich-media content and in structured interactions both human-human and human-machine. These roots endow these capabilities with potentially revolutionary attributes. These attributes include transcending time and space and enabling new forms of scaling. They may enable teaching models to scale to one to thousands at one extreme and 1 to 1 at the other extreme. The technical potential is very real. The infrastructure to deliver is being put in place. How to make it effectively useful to practitioners emerges as the next challenge.

This new class of multiparty, network-mediated, adaptive, interactive, rich-media applications aims at areas such as e-Learning, e-commerce, gaming, and other areas where the quality of the end-user experience is a critical metric of the effectiveness of the application. The spread of these requirements into several areas will drive common functionality - such as adaptation or structured interaction modelling - into mainstream middleware. e-Learning technology then becomes the configuration and structuring of this middleware and related content to select from the repertoire of capabilities those that will provide the needed experience. The focus of recent work has been the generation of specifications for structuring rich-media content for this purpose. We are moving onto research in representing structured, adaptive interactions and various forms of collaboration.

The most challenging work lies in finding means to enable practitioners to make the composition and use of computer-based experiences part of their everyday work. This will require adaptation on their part and a major effort by technologists to bridge the conceptual gap between the technology model and the practitioners' models. Success here has the potential to lead to major transformation, if not revolution, in the delivery of education and training.

The technologists are already working with specialists in pedagogy, with educational policy makers, with behaviourists and other members of the soft sciences. I believe we need to engage with the practitioners, both to understand better how these techniques fit into the practices of training and education today and also to finally bring clarity and focus to my opening question: What critical problems in training and education can we solve through the use of computers?

REFERENCES

Advanced Distributed Learning Network (2002) ADL SCORM specification. [http://www.adlnet.org]

Cetkovic, A., Drexler, H., Guinand, M., Jauslin, D. and Leummens, M. (2002) [http://www.ethworld.ethz.ch/nw/competition/details/?tarnum=3007]

Farrell, R. (2002) Dialogic Engagement for Active Web-based Learning. In *E-Learn 2002 World Conference on E-Learning in Corporate, Government, Healthcare and Higher Education*. Montreal, Canada. In press.

IEEE P1484.12 Learning Object Metadata Working Group.
[http://ltsc.ieee.org/wg12/]

Koper, R. (2001) Educational Modelling Language. Open University of the Netherlands [http://eml.ou.nl/introduction/docs/ped-metamodel.pdf]

Learning Resource Meta-data Specification (2002) IMS Global Learning Consortium. [http://www.imsproject.org/metadata/]

MPEG-7. [http://mpeg.telecomitalialab.com/standards/mpeg-7/mpeg-7.htm]

Sherman, R. A. (2002) Abstraction in Concept Map and Coupled Outline Knowledge Representations. *In Journal of Interactive Learning Research*. In press.
Web Services Description Language (2002) [http://www.w3.org/TR/wsdl]

Knowledge Management in Education

Jane Andersen
Open University, The IT University of Copenhagen, Glentevej 67, DK-2400 Copenhagen NV, Denmark

jane@it-c.dk

Key words: Knowledge Management, Catalyst for Adaptation, Distance Learning, Social Issues

Abstract: When the IT University of Copenhagen started in 1999 there was a manager but there were no students or teachers. In 2002 there are 1200 students and 150 teachers and these numbers will increase. Each teacher is appointed because of his or her outstanding IT knowledge, and he or she chooses the detailed IT content of his or her course. Students, as soon as they register, are given emails and logons to the Intranet, as email and Intranet are two of their main sources of information during their period of study. In terms of integration of IT in teaching the IT University of Copenhagen is similar to most other educational institutions. The knowledge is mainly managed individually by each teacher - tacitly. This paper focuses on the main challenges to successful integration: conditions for generating knowledge on IT integration, knowledge management, management, traditions, social relations and net-based education.

1. INTRODUCTION

This paper is about knowledge management - generation, collection, organisation, dissemination, and usage of knowledge - and the successes, challenges and problems faced when developing and practising knowledge management in education. In my opinion the main aim of knowledge management is the *usage* of the knowledge.

The following reflections are based on the knowledge area 'IT in

Education' with which I have been intensely involved. 'IT in Education' knowledge can typically be categorised at three levels:

- IT in teaching and learning at course level.
- IT strategies at school or institutional level.
- IT at the national level.

2. PERSONAL RELIABILITY

One of the biggest problems is the enormous volume of knowledge about IT in Education, which is not optimally used.

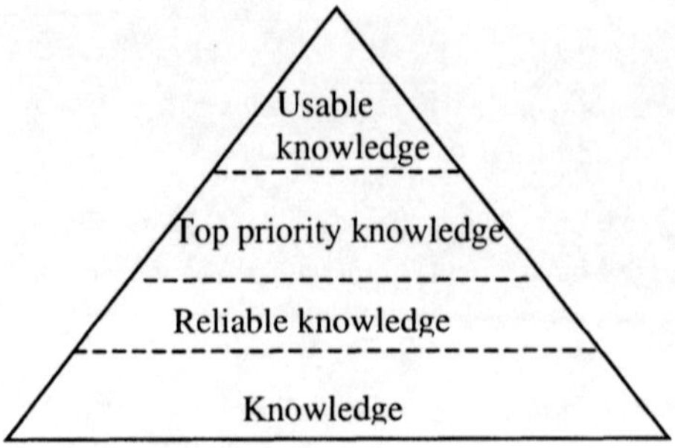

Figure 1. The personal reliability triangle of knowledge management

Postulate 1: I maintain that knowledge is only usable if it is firstly reliable, secondly given priority and finally made usable for the specific user or groups of users (Figure 1). In the education sector important groups of users are the teachers who need knowledge about teaching and learning, the managers who need knowledge at the strategic level and IT project managers and IT responsible staff who need knowledge about teaching and learning as well as knowledge at the strategic level.

Postulate 2: I further maintain that IT can be used for making knowledge usable but human interactions are important and necessary as well.

I will illustrate the importance of the personal reliability triangle through two examples I have experienced. The first example concerns knowledge management at the former CTU, The Danish National Centre of Technology Supported Learning, now known as Learning Lab Denmark. The other example is from my development post at the Open University Department based at the IT University of Copenhagen.

3. CTU

The aims of CTU were to subsidise and support projects aimed at improving educational quality through the adoption of new technologies and to be a national knowledge centre on IT in education.

The main idea underlying the knowledge centre of CTU was that the production of knowledge was effected by the projects. They were paid to produce, structure and report results to CTU. Many Danish educational institutions received experiences with IT and CTU got very detailed knowledge on IT and institutional adaptation. CTU received the knowledge in a printed report and on disk - and we also talked with staff involved in most projects many times during the project period.

The primary way of organising the knowledge was the presentation of project reports on the web. From the projects we also extracted, restructured and concentrated the knowledge and we added services such as a copyright service for usage of electronic educational material. CTU never published a whole project report but extracted knowledge for printed articles and books.

For internal use all information on projects was kept both on paper and electronically. There were three primary ways of disseminating the knowledge: printed, web-based and 'mouth to mouth'.

3.1 Types of knowledge - where were the successes?

We experienced different degrees of success with the different types of knowledge. In respect to reliable knowledge the project groups did their work and returned their knowledge as reports, which went to the web. Some interested people, especially researchers and IT project managers, found these reports and printed them free of charge.

3.2 Problems and challenges

Following interviews with selected project groups and analysis of the project reports specific knowledge was assigned priority by the knowledge centre of CTU. This top priority knowledge was then printed and sold as 'CTU News', or included in books or conferences of anthologies.

Usable knowledge was created by further structuring and concentrating the knowledge and presenting it at thematic conferences, in instructional 'cook books' and as dedicated web services. These ready to use offers of knowledge were avidly requested and sold to many real users who were only interested in the knowledge when it was customised for them.

Unfortunately, not everybody who could potentially gain knowledge by reading and using the resources knew about the books and web-services. This is a challenge for the owners of knowledge as well as the potential users.

4. IT-C

My reflections on the development of net-based education at The IT University of Copenhagen (IT-C) show an example of a development process based on mainly social and ethical decisions.

As head of development at The Open University Department at the three years old IT University my responsibilities are both the strategic implementation at university level and the design of the frameworks for teaching and learning.

At the teaching level one of the biggest challenge is that most teachers are not searching for knowledge on teaching at all (Lorentsen 2000). The teachers are researchers - although they are not researching in education - and outstanding IT experts from industry. Course assessments show that most students are very satisfied with most of the courses.

4.1 Phase 1

The Open University Department started in 2001 with three net-based courses. The three teachers involved in the delivery of net-based education had their own experiences from two years of net-based teaching at other universities. I have never practised net-based teaching, but I have studied and worked with distance education for years in Denmark and internationally (Paulsen 2001).

How could we exchange knowledge? Were the teacher's experiences more or less important than the knowledge I had picked up over the years? I decided on a knowledge exchange model where we talked about important aspects of net-based education at face to face meetings. Knowledge transfer was always based on debate such as "My experiences with ... are ...", and "Experiences from many projects report very good experiences with ... Should we try to invoke this in our concept too?" I categorise this as knowledge exchange by negotiation.

For me the social aspects of the development process were very important. Looking at the speed of development it was not the quickest and most efficient strategy to choose but for developing a "social climate" in our group of teachers I am certain it was the best way.

4.2 Phase 2

The four new teachers joining us in 2002 had no experience of net-based teaching. Each new teacher was introduced individually because I needed to know their teaching background, their experiences with net-based teaching and their feelings about the new way of teaching. Were they expecting net-based teaching to be a challenging friend or a time-consuming enemy?

The knowledge management situation has changed, both because of the new teachers and because of the experiences gathered at the IT-university during 2001. I wrote an eight-page introduction (directives as well as advice and ideas) to net-based education at the university where I collected both experiences from university teaching and my own knowledge in the area. This was the first time that I was able to introduce and implement a more substantial part of my knowledge.

The written introduction was given to all teachers for discussion and stimulation of the exchange of ideas between the 'new' and the 'old' teachers. However, only the new teachers showed up for the meeting I set up for discussion!

I categorise this as knowledge exchange based on the apprenticeship principle.

4.3 Phase 3

At the conclusion of the spring term 2002 we asked the 200 students to evaluate the courses. Their opinions will be very important. It will be my responsibility to make fair conclusions, praise the successes and prepare solutions for adjustments of problematic areas to push the further development process in co-operation with the teachers.

I am sure that some knowledge management can be done by writing a common paper and meeting in groups - but I am also sure that there need to be face to face meetings and talks between each of the teachers and myself.

A further step in knowledge management will be the provision of web-based information for potential students about the qualities of our net-based education. This pre-course preparation and support of students will give them more substantial expectations which everybody involved in net-based education throughout the IT University will be responsible for implementing.

4.4 Characteristics of the knowledge management

Generation and collection of knowledge was carried out by the teachers and myself through years of working experiences before the start of the university. To a great extent the knowledge was organised as tacit knowledge in our heads in phase one. This organisation of knowledge changed during phase two where I evaluated the experiences and made priorities before making dedicated directives and catalogues of ideas about net-based education for teachers as well as for students. The knowledge is disseminated through talk and print as well as by email and Internet web-sites. Usage of knowledge could be seen during the term where some students used the "good old" question and answer method before searching through the directives while other students did the opposite.

5. IT IN KNOWLEDGE MANAGEMENT

What can IT do for knowledge management in the education sector? IT cannot make knowledge reliable or give priority to knowledge, but IT can be used for making knowledge usable. As usable knowledge is the most important form of knowledge IT is very important for knowledge management in the education sector.

An outstanding example of usable knowledge is the Danish Poseidon Project (Poseidon 2002), which has a web-based toolbox for development of IT strategies at educational institutions. Through pre-prepared checklists and questionnaires the educational institution completes a statement of current practices and is helped to articulate their vision, aim and plan of operation of ICT integration. The toolbox has been tested and used by several hundred Danish educational institutions from primary schools to universities. It has been translated into English, French, Portuguese and Dutch as a part of the European Network of Innovative Schools (ENIS 2002) and is freely available on the web.

5.1 Challenges and real problems

Working with IT in education for many years I have identified the following challenges and problems which make it complicated to disseminate knowledge and secure usage of knowledge.

Usable knowledge is individually identified:

- You can only expect usage of knowledge if the user thinks that the knowledge is important.
- Teachers, IT project managers and managers expect different kinds of usable knowledge.
- Researchers focus on reliable knowledge, developers on high priority knowledge and users on usable knowledge.

Most knowledge about IT in education is never made usable:

- Many projects have produced reliable knowledge on IT in education, but the investment in the process of making reliable knowledge usable is too small.

The characteristics of usable knowledge need more investigation.I am finding these problems formidable and my suggestion is that all universities and other educational institutions, together with regional and national knowledge centres, should set up strategies for all aspects of knowledge management.

6. EDUINFOLEARNKNOW

The future knowledge management and learning situation will be existentially affected by the ongoing mix of education, information, learning and knowledge management into a kind of EduInfoLearnKnow (Figure 2).

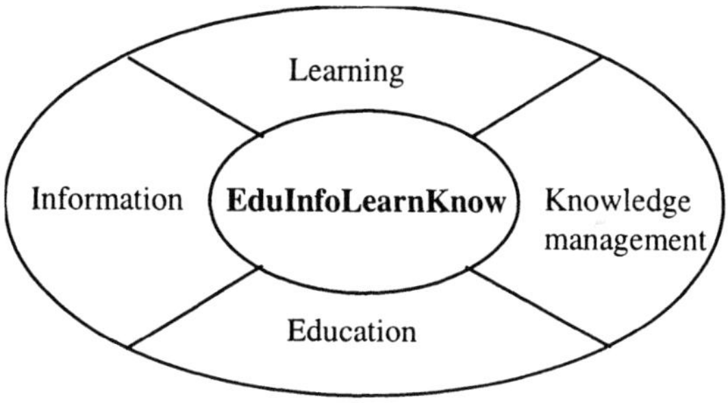

Figure 2. EduInfoLearnKnow

A tendency at the moment is to label everything in this area e-learning, which definitely will change e-learning and the whole principle of learning! The last time I met with industry representatives they told me that it makes no difference how and from where they get knowledge - as long as they are gaining something.

If we don't react I think e-learning will very shortly end up as e-information. That is not the result of IT in education I am looking forward to!

REFERENCES

Lorentsen, A. (2000) *Aspekter af teknologistoettet fjernundervisning paa universitetsniveau.* University of Aalborg.

[http://www.sprog.auc.dk/persons/]

Paulsen, M. F. (2001) *Netbasert Utdanning*. Gyldendal. Denmark. (Danish introduction by Andersen, J and Andresen, B. B.)

The IT University of Copenhagen (2002).

[www.it-c.dk]

Poseidon (2002). The Danish Poseidon Project.

[www.poseidon.uvm.dk]

CTU (2002) The Danish National Centre of Technology Supported Learning.

[www.ctu.dk]

LLD (2002) Learning Lab Denmark.

[www.lld.dk]

ENIS (2002) European Network of Innovative Schools.

[http://www.enis.eun.org/Toolbox/overblik.htm]

Learning and Teaching in Socio-technical Environments

Thomas Herrmann
Informatics & Society, University of Dortmund, D-44221 Dortmund, Germany

thomas.herrmann@udo.edu

Key words: Collaborative Learning, Communication, Learning Environments

Abstract: Computer supported learning environments can be analysed as socio-technical systems. New aspects of system theory, such as self-reference, autopoiesis, autonomy and contingency can be correlated with concepts of the research on social capital and social identity. This provides a basis to interpret empirical findings in the case of the practical usage of a learning environment and to propose measures of improvement. Ubiquitous self-description and continuous facilitation of communication processes can be considered as important success factors.

1. INTRODUCTION

The usage of information technology for teaching and learning not only supports these activities but also influences and forms their characteristics. In an increasing number of cases, learning processes and technical systems not only have a co-appearance but are highly interrelated. The result is a specific kind of socio-technical learning and teaching systems which have to be carefully understood to achieve a successful adaptation between organisational processes and technical design. In universities, various kinds of technical systems have to be taken into consideration such as interactive e-learning modules covering specific topics, platforms employed to store and provide access to traditional learning material via WWW, systems supporting collaborative learning, and applications which facilitate organisational tasks.These systems are interwoven with lectures, practising groups and seminars. Since these different types of technical support will be increasingly combined and embedded into appropriate organisational processes and didactical methods of teaching and learning, we assume that universities will become socio-technical systems where the use of technology is directly related to the characteristics of teaching and learning and vice versa.

This paper describes the key concepts which theoretically characterise socio-technical systems and are practically relevant for teaching and learning. The relevance of social identity and social capital for the development of communities as a success factor of computer supported collaborative learning is emphasised. The influence of these concepts for the development and use of technical systems in universities is illustrated through consideration of the collaborative learning environment KOLUMBUS. This not only covers a technical system but also organisational procedures and rules which are relevant to support collaborative learning.

The relevance of a socio-technical perspective is outlined by referring to empirical findings about the usage of KOLUMBUS. While KOLUMBUS is an example on the level of seminars, we conclude with a section which transposes our considerations and requirements on the level of the university as a whole.

2. THEORY OF SOCIO-TECHNICAL SYSTEMS

2.1 The terminology

The term socio-technical system is often used in its literal meaning simply referring to a combination of organisational, technical, educational and cultural structures and interactions. An historical consideration reveals the potential and the deficits of this approach. The term is linked to a tradition founded at the Tavistock Institute in the 1950s: Trist, Bamforth and their colleagues analysed the situation in the British coal mining industry where productivity had dropped after the introduction of new technology (Trist and Bamforth 1951). Their findings led them to conclude that the productivity of a working organisation not only depends on social aspects of work organisation but also on the well considered integration of the technical systems being used. They subsequently developed an elaborated method of analysis and intervention in working organisations - the so called socio-technical approach. This was adopted and further developed by Emery and

Thorsrud in the Norwegian industrial democracy project (Emery and Thorsud 1976).

In the 1960s and later it became an important aspect of their work to face the theory and practice of management with the relevance of the social aspects of work organisations and the fact that human co-operation cannot be controlled in the same way as machinery. It was only later that the socio-technical approach was transferred into the area of computer system application - for example by Enid Mumford (1995). She claims that the aim of socio-technical design is to give equal weight to social and technical issues when new work systems are being designed (Mumford 2000). Others, such as Eason (1988) also emphasised the aspect of organisational change. He states that a socio-technical system can be a subject of design activities and that the starting point of this design has to be a careful definition and analysis of the social system including people's work roles and their ways of co-operating.

These socio-technical approaches, as well as other concepts of systems thinking which take "soft" aspects into account (Checkland 1981), are all oriented towards open systems. The behaviour of open systems is influenced from outside and depends on the stimuli which occur in their environment. This focus implies several deficits. Most importantly it cannot explain the lack of causality between influences from outside and the reactions of a human individual or a social system, such as a company, a department or a class of students. The same deficit applies to the relationship between social systems which *produce* technology and those which *use* it. The interaction between processes of producing, introducing and using technology does not follow the pattern of cause and effect.

2.2 New approaches in system theory

To overcome these deficits it is sensible to refer to more recent concepts of systems theory developed with respect to living systems (Maturana and Varela 1987), self-referential and self-organising systems (von Foerster 1970), and neo-constructivism (von Glaserfeld 1984). Luhmann (1985) combined and integrated these concepts into a theory of social systems. This theory helps improve our understanding of socio-technical systems and our appreciation of the difference between social and technical structures.

The main difference is that cognitive or social systems are not only open - so they can interact with their environment - but are also closed. Thus the final selections of the system's behaviour depend exclusively on its own structure (Varela 1981). The reactions to influences from outside and the way the perceived behaviour of its environment is transformed into information processing or into operations are determined by the system's own structure and not from outside. This phenomenon of referring exclusively to itself is described as self-reference (Luhmann 1993). It enables a system to behave autonomously. Autonomy does not imply that it is not related to its environment: the impulses from outside can initiate the selection between different alternatives. Typical examples are communication or teaching processes: a message in the course of a communicative process establishes a scope of expectable reactions on the part of the recipient, however the selection of the reaction depends exclusively on the recipient. This relationship between a message and the following reaction - for instance a question and an answer - can be described as "contingent" (Luhmann 1995; Parsons 1967). Contingency applies also to the relationship between the efforts of a teacher on the one hand and the development of a student's competence on the other.

The phenomenon that living, cognitive or social systems are self-dependent or self-referential is not self-evident since they could also be considered as a product of their environment. Maturana (1985) explains the possibility of self-reference by referring to the observation that living systems constantly produce themselves. This structural characteristic is called "autopoiesis": a network of interacting elements constantly produces the elements and interactions of which this network is built (Maturana 1985). An autopoietic system determines by itself which elements do or do not belong to it. Autopoiesis includes a continuous maintaining of the system's identity and its boundaries with the environment (Varela 1981). In the course of autopoiesis, the systems establish and reproduce their own rules of their functioning as well as the way they perceive and react to their environment. These thoughts also form the basis of constructivism (von Glaserfeld 1984) - how reality is perceived depends on the structure of the observing system and there are as many realities as there are observing systems. The concept of constructivism has increasing relevance for the analysis of teaching and learning processes and is related to a learning paradigm with two interesting characteristics:

> "learning is an active process of constructing rather than acquiring knowledge and instruction is a process of supporting that construction rather than communicating knowledge." (Duffy and Cunningham 1996)

In contrast, the characteristics of technology must not be described by referring to contingency, self-reference or autopoiesis. A technical system is not a product of its own but is made and controlled from outside. Thus, it is determined from outside whether an element is part of the system or not. Technical systems serve purposes which do not lie within themselves but are assigned from other systems. The successful usage of technology requires pre-defined and reliable input-output relations instead of a contingent relationship between the user's action and the technical system's reaction. Even in the case of so called "learning machines" a program determines how the machine learns and what kind of rules about the structure of its environment it can develop.

In the case of socio-technical systems, these two types of fundamentally different systems are combined and integrated. Technical systems are always a part of the environment of a social system, but, in the case of socio-technical systems, the interaction with the technical system is not coincidental but is systematically reproduced. We assume, in general and especially with respect to collaborative learning systems, that the following characteristics are decisive for a socio-technical system.

- Reciprocal indispensability: The technical system is as indispensable for the behaviour, the interactions and the autopoietic process of the socio-technical system as the communication processes of the social system.
- Reciprocal forming: The possible input-output relationships of the technical system have a forming influence on the communication structures of the social system and vice versa. Technical elements are an integral part of the socio-technical system. Their absence would affect the characteristics of the social system and would alter the process of its constant renewal.
- Ubiquitous self-description: The socio-technical system includes processes of describing itself. This self-description does not only identify the communication acts that make up the social system but also identifies relevant technical elements as integral parts. Self-description is partially represented as semantic artefacts which can also be found as representations displayed by the technical system.

2.3 Self-description

All three of these characteristics are indispensable. We consider the phenomenon of self-description needs special attention since it is not only an analytical category but also a matter of design. Luhmann (1995) states that social systems are self-referential and therefore include descriptions which are necessary to guide the continuous process of self-remaking or self-renewal. Self-descriptions can be found in different forms: as part of temporal communicative articulations and, consequently, as persistent semantic artefacts. Socio-technical systems can already incorporate several types of descriptions. On the one hand, the description of social-systems is encoded in conventions, in the verbal elements of a shared semantic system, a meaning system or even in written rules or laws. In organisational systems, co-ordination is a major issue that is dealt with by these descriptions. They are also relevant to co-ordinate processes of collaborative learning. Technical systems are described by engineering artefacts. Computer systems in particular are described by software engineering oriented modelling methods or program code. Today both types of descriptions are separated, often not made explicit and hardly reflect each other. It is a matter of design to offer methods which support two requirements:

- The facilitation of integrated descriptions covering both technical and social aspects.
- The option to make these descriptions part of communication processes or technically stored and displayed representations.

The modelling method SeeMe (semi-structured, socio-technical modelling method) (Herrmann and Loser 1999), meets both requirements since it can handle contingency, incompleteness, and freedom of decision as well as well defined input-output relations.

We assume that self-description is of high relevance for the success of computer supported collaborative learning since the students need a basis to understand themselves as a learning community which has a certain social identity - even if their interactions are partially computer-mediated. The mediation offered by computers can both support and hinder the development of a (virtual) community's social identity. To obtain a better understanding of the relevance of social identity and its pre-conditions, it is important to refer to its relationship with the concepts of social capital (Putnam 2000; Fukuyama 1999) and communities of practice (Wenger 2000).

2.4 Self-reference in the context of social capital and communities of practice

We assume that collaborative learning cannot evolve and succeed without the building of a network, norms and trust - exactly the features of an organisation which are the essence of Putnam's social capital since they facilitate co-ordination and co-operation for mutual benefit. According to Wenger (2000) social capital as a result of trust is the basis for the development of communities. He connects the concept of social capital with the concept of social-identity, which he considers is indispensable for the existence of a community. The relevance of social-identity can be confirmed from the perspective of the theory of social systems (Luhmann 1995): social systems maintain their identity and their borders towards the environment by being self-referential. Wenger adds practical insights to this theoretical approach by describing how social-identity can be achieved or maintained: mutually experienced events, rituals and conventions contribute to social-identity as well as the engagement and the alignment of a community's members and the evolution of their shared meaning system. We suggest self-description can also contribute to social-identity and illustrate this in the context of the collaborative learning environment KOLUMBUS.

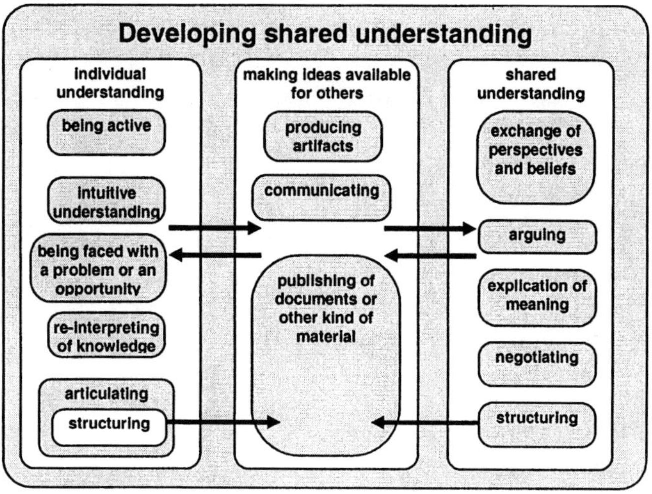

Figure 1. The interplay between individual and shared understanding

Social capital has different dimensions (van der Speck et al 2002) which can also be decisive for the success of collaborative learning:

- The structural dimension (Edvinson and Malone 1997): who can be reached, how and when.
- The relational dimension (Fukuyama 1999): personal and emotional bonding to one another.
- A cognitive dimension (Brown and Duguid): the shared understanding.

In the context of analysing and supporting collaborative learning, the relationship between shared understanding and shared meaning systems on the one hand and social capital and social-identity on the other is extremely relevant. This relationship influences the conditions for bridging the gap between individual understanding and shared understanding. Figure 1 describes some details of the interplay between individual and shared understanding and can serve as a basis for designing collaborative learning environments. Individual understanding is accompanied by human activities and is intuitive until the individual is confronted with a certain problem or an unusual opportunity. Both may lead to a re-interpretation of knowledge and to an articulation of parts of the knowledge - either to support a person's own reflection or the interaction with others. The interaction represents an attempt to achieve a better understanding in the course of re-interpreting knowledge. It is mostly based on communication but can include the production of artefacts or the publishing of documents which do not need a pre-specified recipient. Making ideas available to others implies that differences may become apparent. This leads to an exchange and negotiation of perspectives, believes or arguments. Even the meaning system used for the discourse, can become a matter of re-interpretation and negotiation. All the activities in the diagram are relevant for the development of shared understanding. They have to be supported in a computer supported collaborative learning environment. In the context of computer mediation, the exchange of documents and the possibilities for giving them a meaningful structure gain relevance.

3. DESIGNING SOCIO-TECHNICAL SYSTEMS FOR COLLABORATIVE LEARNING

3.1 KOLUMBUS as an example for designing socio-technical systems

The collaborative learning environment KOLUMBUS was developed to support students during their work to achieve mutual results and shared understanding. It is a task-oriented system. The task is decisive. It should initiate a process of investigation and research and support the growth of shared understanding between students. We wished to offer a socio-technical system which supports the integration of collaborative research and learning. Therefore a technical system KOLUMBUS and a collaborative learning process "working with KOLUMBUS" were developed. They are described in "KOLUMBUS: Context-oriented communication support in a collaborative learning environment" later in this book. The core concept is to support articulation by offering the possibility to add annotations to elementary items of text or multi-media material. The annotations can be extended to a discussion thread and the recipients of communication acts can be selected by flexibly determining the access rights. For instance, a student can control whether he or she alone can read an annotation or whether the teacher or a selected group of students can read it.

KOLOMBUS was tested in a real seminar which is a mandatory part of the computer science program at the University of Dortmund. The task was to prepare a talk and a presentation to be given at a meeting of all 16 seminar participants. The students had to support each other by commenting their preparation material and to avoid overlapping between the different talks. Some of the empirical findings can be directly related to characteristics of socio-technical systems and the underlying theoretical concepts. The experience with KOLUMBUS clearly shows the relevance of the aspects of socio-technical systems for the development of (collaborative) learning environments, particularly "supporting self description" and "(communicative) facilitation".

3.2 Supporting self-description

The model of the *KOLOMBUS-process* is a semantic artefact representing a part of the socio-technical system and thereby supporting its self-description. A derivative of this process model was used to instruct the students and to explain to them how the seminar is organised, what they were expected to do, and how the technical features of KOLUMBUS should be used.

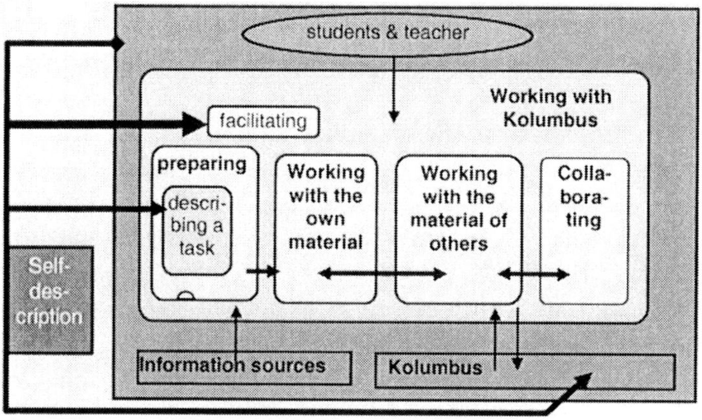

Figure 2. The description of the KOLOMBUS-process being a part of the process itself

Thus, a representation of the *KOLOMBUS-process* was part of the training and preparation of the seminar - both activities are elements of this process. This is a typical self-referential constellation. Furthermore, the model of the *KOLOMBUS-process* was not only a part of the communication course of the preparation but was also stored as a document in the KOLUMBUS database. This document was potentially present and usable for the participants during the whole *KOLOMBUS-process* and was available to be part of self-descriptive processes within the socio-technical system. In Figure 2 a simplified process model describes how a representation of the *KOLOMBUS-process* can be considered an entity which is an integral element of certain activities and the technical system itself. Beside the representation of the process, it is also crucial that the description of the underlying task (as the basis of the learning process) is permanently accessible for the participants. Therefore, the task outline also serves as a self-description. It is available as content which is stored in the system and can be the subject of annotations.

Not only the abstract but also the concrete processes of their research and collaboration should be comprehensible for the participants. How an idea, a solution, or a discourse has evolved or how a mutual result has been achieved should all be documented. Additionally, the contributions and the interests of others and the development of their knowledge should be optionally comprehensible with the help of the system. This can be supported by awareness functionalities (Sohlenkamp and Chwelos 1994). They make the actions of others in computer-mediated environments more visible. Awareness of others' behaviour and expectations can be considered a pre-condition of the evolution of social relationships, social-identity and social capital.

3.3 Relevance of (communicative) facilitation

It is an important but not a surprising finding that the availability of a self-description document is not sufficient to guide the process of collaborative learning. A facilitation of the communication processes is needed to permanently give impulses and to focus the attention on the requirements of the *KOLOMBUS-process*. For example, the students had to be challenged to start with commenting on what others had written. The intended process has to be explained continuously. A facilitator has to give hints - what can happen next or on which contributions someone should react. The activity of facilitating is part of the collaborative process and therefore part of the *KOLOMBUS-process*. Consequently, it is part of the self-description and it uses this description by itself (Figure 2). During this process the interaction between students has to be stimulated. The availability of learning material and communication support on the World Wide Web makes the vision of learning-every-time-and-everywhere possible but by no means guarantees it is brought to reality.

The experience reveals that web-wide availability and ubiquitous opportunities to participate in the system are not sufficient. A facilitator has to provide well-directed stimulation and to propose phases of synchronisation during which the participants should interact concurrently. Furthermore, phases of intensive interaction and face-to-face communication are needed to promote the social relationships and to help to establish social capital. Both the teachers and students should have the possibility to play the role of a facilitator. This implies that they also have the opportunity to alter or adapt the *KOLOMBUS-process*. Consequently, they can consider their own interactions as elements which shape the system and contribute to its permanent renewal in the sense of autopoiesis. The possibility of this kind of system adaptation can be a pre-condition for the adoption of the collaborative learning environment. The students have the opportunity to feel an integral part of a socio-technical teaching and learning system.

4. CONCLUSION: THE UNIVERSITY AS A SOCIO-TECHNICAL SYSTEM

The findings and principles, which refer to socio-technical systems, may be applied not only on the level of collaborative learning environments but also in the context of larger units. Increasingly, the university as a whole can be considered as a socio-technical system. Also on this level, appropriate organisational processes and technical means should improve the integration of the activities of learning, research and teaching. The organisational processes correspond with administration tasks such as configuring the access rights for users, procedures of registration, testing or assessment, and establishing services for multimedia support. A university has, on the one hand, to technically improve the possibilities for mobility and ubiquitous learning which is independent from limits of time and space. On the other hand it has to establish organisational procedures which guarantee that the members of the university have sufficient opportunities - by direct communication or meetings - to develop community-oriented social-identity and to build social capital. This needs facilitation, processes of mentoring, problem and project-orientated teaching, and continuous stimulation of communicative exchange. It also requires a proper integration of spatial and electronic functionality. Multimedia technology should not exclusively be available at isolated workplaces but should also be an integrated part of meeting facilities.

A university can be considered as a socio-technical system which consists of many sub-systems. All the departments, groups, centres or temporally built communities can also share the properties of socio-technical systems. Appropriate organisational procedures, training opportunities and technical means have to ensure that innovative solutions for teaching and learning are deployed to all these sub-units and that the process of innovation becomes continuous and sustainable. For instance, an e-learning module should not only be tested in one setting for one time, but should be adopted for many purposes if the test proves successful. This goal requires flexibility and adaptation. The process of integrating and adopting such modules in different sub-units of a university needs professional support and facilitation.

In the area of technical learning and teaching support, a university needs standards so that it can establish as a socio-technical system as a whole. On the other hand, flexibility is required to support the special needs of its subunits. The balance between standardisation and flexibility is itself a matter of well integrated and adapted socio-technical processes. It has to share the main characteristics of indispensability of its components, reciprocal forming and ubiquitous self-description. Self-description on different levels supports self-reference, sustainability and social identity. It has to be completed by manifold forms of facilitated communication to initiate processes of change and to support the building of social capital.

Acknowledgements

This paper became possible thanks to the work of the members of my group, especially Andrea Kienle who has - together with Carsten Ritterskamp - accompanied the development of Kolumbus and conducted the empirical investigations, and Gabriele Kunau who has contributed important details to my knowledge about the theory of socio-technical systems and self-description. My understanding of "shared understanding" was decisively influenced by Gerry Stahl during his time as a research professor at the University of Colorado, Boulder. Etienne Wenger's talk and his availability for intensive discussion at the Workshop "Social capital and IT" were a basis for my insights into the relationship between self-reference and social identity. This workshop was organized by Marlene Huysman and Volker Wulf who thereby have provided an excellent opportunity to connect my knowledge about social systems theory with the social capital approach.

REFERENCES

Brown, J. S. and Duguid, P. (2000) *The social Life of Information*. Harvard, Boston.

Checkland, P. (1981) *Systems Thinking, Systems Practice*. John Wiley and Sons, New York.

Duffy, T. M. and Cunningham, D. J. (1996) Constructivism: Implications for the Design and Delivery of Instruction. In *Handbook of research for educational communications and technology*. D. H. Jonassen. Macmillan, New York. pp. 170-198.

Eason, K. (1988) *Information Technology and Organisational Change*. Taylor and Francis.

Edvinson, L. and Malone, M. S. (1997) *Intellectual Capital*. HarperCollins Publishers.

Emery, F. and Thorsrud, E. (1976) *Democracy at work*. Martinus Nijhoff Social Sciences Division, Leiden.

von Glasersfeld, E. (1984) An introduction to radical Constructivism. In *The Invention of Reality*. P. Watzlawick. New York. pp 17-40.

von Foerster, H. (1970) Thoughts and Notes on Cognition. In *Cognition. A multiple View*. P. L. Garvin, (ed.). New York. pp. 25-48.

Fukuyama, F (1999) *Social capital and Civil Society.* The Institute of Public Policy, George Mason University, October. Prepared for delivery to the IMF Conference on Second Generation Reforms.

Herrmann, T. and Loser, K-U. (1999) Vagueness in models of socio-technical systems. In *Behaviour & Information Technology*: Special Issue on Analysis of Cooperation and Communication 18(5), pp. 313-323.

Luhmann, N. (1993) *Essays on Self-Reference.* Columbia, New York.

Luhmann, N. (1995) *Social systems.* Stanford

Maturana, H. (1980) Man and Society. In *Autopoiesis, Communication and Society*: The theory of autopoietic system in the social sciences. F. Benseler, P. Hejl,and W. K. Köck, (eds.). Frankfurt. pp. 1-31.

Maturana, H. and Varela, F. (1987) *The Tree of Knowledge: The biological roots of human understanding.* New Science Library, Boston, MA.

Mumford, E. (1995) *Effective Systems Design and Requirements and Analysis - the ETHICS approach.*

Mumford, E. (2000) A Socio-Technical Approach to Systems Design. In *Requirements Engineering* (2000) 5. Springer, London. pp. 125-133.

Parsons and Talcott (1967) *The social System.* Lowe and Brydone Ltd., London.

Putnam, R. (2000) *Bowling alone: The collapse and revival of American Community.* Simon & Schuster, New York.

Sohlenkamp, M. and G. Chwelos (1994) Integrating Communication, Cooperation and Awareness: The DIVA Virtual Office Environment. In *Proc. of Conference on Computer Supported Cooperative Work,* Chapel Hill, NC, USA, R. Furuta and C. Neuwirth (eds.). ACM Press. pp. 331-344.

van der Spek, J., Mulder, I., de Poot, H. and Moelaart, F. (2002) A comparative Study on social capital in communities in three sectors. In: Research Workshop on Social Capital and IT. May 2002. M. Huysman and V. Wulf (eds.). Unpublished.

Trist R. E. and Bamforth, K. (1951) Some social and psychological consequences of the long wall method of coal getting. *Human Relations* 4, pp. 3-38.

Varela, F. J. (1981) Autonomy and Autopoiesis. In *Self-organizing Systems: an interdisciplinary approach.* G. Roth and H. Schwegler (eds.). Frankfurt - New York. pp. 14-23.

Wenger, E. (2000) *Communities of Practice.* Cambridge University Press, Cambridge.

Working Group Reports

Key words: Collaborative Learning, Cyber Risk, Digital Divide, e-learning, e-literacy, Equity, ICT, Imported Culture, Learning Environments, Power Shift, Privacy, Security, Social Issues, Virtual Reality

Abstract: These are the nine edited reports prepared by the different Professional Working Groups at the SECIII conference. Each represents many hours of discussion - both during designated conference discussion time and at more informal sessions. All of these reports were delivered at the final session of the conference

1. E-LITERACY

Just the term E-literacy raised a host of questions and generated a tremendous amount of discussion. What is the problem with E-literacy - is it the 'E' or the 'literacy'? Does 'E' refer to keyboarding, the content, or the interface? Is E-literacy just about using a computer and knowing what the software is used for? At what level of use are we talking? With reference to the 'literacy', do we mean 'competence' and if so, in what? Is this 'using' a computer? Is E-literacy a sub-domain of general literacy? What are the parallels between E-literacy and literacy and numeracy? Can a good, or acceptable working definition of literacy, numeracy and E-literacy be described as "to be on familiar terms with ....?.".

The group felt the term E-literacy could be given a more precise meaning by adopting a structure proposed by Heinz Moser for Media Competence. Media Competence involves - Technical aspects, Cultural aspects (such as codes and forms of communication), Social competence (i.e. people should be able with using and not using ICT) and a Reflective aspect (they should reflect about ICT in society).

It may well be that there is no problem for the majority of students who live in an E-rich environment: they can acquire the competencies they need rapidly and easily. However, what about the 10 - 20% of children who - according to David Wood in his keynote address - find it difficult, even impossible, to learn using computers? It is clear that, initially at least, it is the teacher whose development has to be supported. This involves providing time, opportunity and training.

We have to discuss innovation with teachers and encourage them to combine teaching and learning methods, which should be offered in an ICT-rich environment, within which pupils will be enabled to make choices. As part of their general, out-of-school E-literacy pupils spend a lot of time playing computer games - as mentioned in the keynote by Colin Harrison. We have to find ways of using such games for 'educational' purposes.

Our conclusion is therefore that we do not need to teach E-literacy, but must provide plentiful and 'sensible' opportunities for both students and teachers to explore and resolve problems and issues using E-literacy and so acquire particular E-literacy skills, techniques and understanding. However, this implies changes in the way student learning is organised, so that participants in the student's education (parents, pupils, teachers) are prepared to adopt different roles at various times. Finally there are some basic concepts in E-literacy that have to be learned by all pupils and teachers.

Acknowledgement

The above summary is based on a report of the SECIII Working Group on E-literacy. Participants were: Yvonne Büttner (Chair), Barry Blakeley (Rapporteur), Beat Döbeli, Dirk Draheim, Michael Janneck, Peter Micheuz, Robert Munro, Prathiba Nagabhushan, Clara Oliveira, Ulrik Schroeder, Alnaaz Kassam, Tom van Weert.

2. COLLABORATIVE LEARNING

Collaboration intensifies the human aspect of learning. It increases our learning potential and empowers us with the knowledge of others. It is an ancient educational model (knowledge building in pairs by Plato and Socrates, study groups at medieval noble courts and constructive learning of the Reform Pedagogy movement).

Collaborative activities are democratic by nature, but require careful planning, monitoring and scaffolding by the teacher. Collaboration makes group members more flexible and eventually replaceable as the group as a learning community shares knowledge and experiences.

Concepts of collaboration vary from country to country – some projects are more mentoring-intensive, others aim at substituting teacher work through student engagement. In order to develop successful strategies and teaching aids, mental models and learning theories have to be used. Some important contributions in this area include Kenneth Brophy's Collaborative Learning Theory, Action Learning Theory by Ray Revans and Entrainement Mental by Jean Francois Chosson.

The organisation of knowledge in collaborative learning projects is crucial: currently we see rigidly controlled input and output methods and linear curriculum structures, which do not generally support the collaboration of learners. In a collaborative learning environment students should acquire *co-operative behaviour and learning content* at the same time.

ICT can support but may also hinder collaboration. All ICT-based collaborative projects should include face-to-face components to motivate learners and increase their social skills as well as their learning skills. Important bodies of research support the use of collaborative learning through ICT at all levels of education - such as:

- CSCL environments and cognitive tools e.g. Knowledge Forum by M. Scardamalia and K. Bereiter, University of Toronto.
- International digital projects of the European School Net (EUN).
- SEMIK Project which is concerned with collaborative learning experiments, created by H. Mandl at the University of Munich.
- KOLUMBUS Learning Space developed by the University of Dortmund.
- Collaborative mentoring at La Villa Media, Grenoble.
- The Jigsaw Classroom Method proposed by Aronson.

Collaborative learning methods are generally *taught but not experienced directly* as part of Theory and Practice of Learning studies in teacher training. Theoretical knowledge is not applicable in classroom practice unless teaching methods are associated with this knowledge.

Assessment Issues

Collaborative learning requires suitable assessment methods that value teamwork. If individual achievement is the only valuable outcome, group members will compete instead of supporting each other. Traditional examination and grading systems discourage teamwork.

Research at Budapest University on personality traits and work in CSLEs has found that the personality of the learner and teacher contribute hugely to the success or failure of participation in a collaborative learning experience. Some characteristics are modifiable through successful mentoring but certain learning and teaching styles are not. In Australia the "Electronic Peer Review" method under development at the University of Adelaide and the computer-based essay evaluation software used at the University of Melbourne offer promising assessment alternatives.

School disciplines or learning areas allow for different collaborative methods. Some knowledge items, skills and abilities can be fostered through collaboration while others require frontal or individualised approaches.

ICT can support accurate and detailed assessment of the collaborative process, but privacy needs to be protected and too much control avoided. Some educators believe that the collaborative process should not be monitored and only the outcomes should be assessed – others find ICT-based evaluation of collaborative learning important for developing successful teaching strategies.

Ideas for Action

Collaboration as a learning activity should be encouraged at all levels of education and professional life. Present assessment practices - both at school and in professional life - do not support co-operation.

Mental models of learners and *learning models* should be conceptualised before planning a collaborative learning project.

Human aspects of collaboration in an ICT-supported environment should be carefully studied and the results used in CSLE design.

Widely used ICT-based learning tools and environments should be assessed to find out if and how they support collaborative learning and other contemporary educational paradigms.

Collaboration needs a supportive institutional environment for effective implementation. *ICT acts as a catalyst, evaluator and moderator* for collaborative learning and makes desirable communication, mentoring and co-operation practices easier to realise with large groups of learners. Face-to-face teaching and tutoring is and will always be, however, at the core of education.

Acknowledgement

The above is based on the report of the SECIII Working Group on Collaborative learning. Participants were: Bernard Cornu (Chair), Andrea Karpati and Anna Strehler (Rapporteurs), Jane Andersen, Iolanda Cortelazzo, Dirk Draheim, Ruth Messner, Guido Rößling, Sjoerd de Vries.

3. SOCIAL ISSUES AND THE SHIFT OF POWER

Looking at the social issues associated with Information and Communication Technology (ICT) the following observations can be made:

- Business culture is increasingly competitive.
- It is difficult to work and live in a 'Tree' structure and a 'Network' structure at the same time.

- Digitalisation is the "most extreme" form of abstraction.
- Internet is transforming social interaction among different age groups in society in all countries.
- Many university students in Europe are dissatisfied with their curricula.
- ICT is a major component in merging personal, private, leisure and work time creating increased stress.
- In contradiction to claims made in the nineties of the last century, ICT has not decreased the gap between North and South - it has worsened the situation.
- The frontier between fact and fiction is fuzzy.
- Important areas of ICT "know-how" still lack understanding on a conceptual level.
- Today we have more and more information but less and less critical understanding.

We would therefore like to recommend the following:

- Emphasise critical analysis and become more a producer than a consumer.
- However, do not apply the rules of the production process to the learning process.
- Be aware of different rules and regulations in different kinds of environments (school, job, games).
- Foster interpersonal and social relationships among individuals and/or groups in a society/countries.
- Change learning from a push model to a pull model.
- Take into account and integrate many power shifts like: Teaching to learning; global to local; central to peripheral; traditional to new media; private to public; school to society; individual to groups (Increase the multiplicity of partners).
- Try to integrate ICT into society and not vice versa.

Concrete actions which can be undertaken in view of these recommendations are:

- Do not be afraid to tell students that you do not fully understand all of the leading edge concepts.
- Provoke and facilitate frequent face-to-face exchanges and communication among top professionals in industry, top academic researchers and decision makers in education.
- Stimulate young researchers to carry out research into social issues/power shifts.
- Establish links with the concrete actions of the other working groups in respect of equity, imported culture and cyber crime.

- Include social aspects/power shift issues in teacher education in a comprehensive manner.
- Help parents understand the importance of social aspects in child development.
- Take the opportunities offered by ICT to allow young people to teach adults about technology (use of a mobile, sending an email, programming a DVD recorder) to overcome the generation gap.
- Focus on learning by doing - become a coach or a mentor for others.
- Explore and provide the possibility of 'Time outs' for personal development during the life long learning process.
- Encourage each school actor to be a producer by initially focusing on the process of learning and not on the end product. Increasingly refining the product comes later.

Acknowledgement

The above is based on the report of the SECIII Working Group on Social issues and the shift of power. Participants were: Raymond Morel (Chair), Muddassar Farooq (Rapporteur), Dirk Draheim, Marc Pilloud.

4. VIRTUAL REALITY

The following observations can be made about Virtual Reality:

- Virtual Reality allows you to move through a world without limits.
- The more senses are involved the more the experience is Virtual Reality.
- It is learning by experience (but there can be risks involved).
- It allows sharing of experiences.
- It enables financially efficient and time efficient creativity.

The following recommendations with respect to Virtual Reality are proposed:

- Explore historical environments.
- Develop Virtual Reality experiences for special needs students.
- The educational community should share information about Virtual Reality.
- One set of Virtual Reality equipment should be provided for each school.
- Develop good models for educational use.
- At future IFIP events a 'track' on Virtual Reality should be established.

Acknowledgement

The above is based on the report of the SECIII Working Group on Virtual Reality. Participants were: Pieter Hogenbirk (Chair and Rappporteur), Maria Iding, Wolfgang Luther, David Passig, Marc Pilloud.

5. EQUITY AND THE DIGITAL DIVIDE

The introduction of Information and Communication Technology produces Haves and Have-Nots due to the following:

- Technical aspects of infrastructure, ability and bandwidth: there are countries which do not possess the technical infrastructure to support this technology – sometimes there is not even electricity; Even if there is access to computers and electricity there might be further limitations, such as in bandwidth.
- Gender: fewer girls are involved in ICT; is this because there is not a conducive environment for women and girls?
- Politics: ICT is a threat for countries which do not want the free flow of information.
- Techno phobia; there are technophobes who are scared of the technology and will not use it even if they have access.
- Language: a dominance of English and Western language which limits access to information.
- Literacy: access to computers and the Internet is seriously limited when users are illiterate. However, people may also have no 'lectronacy', which means they are illiterate with respect to electronics.

There are, however, positive points:

- Technical: sometimes the need for ICT allows certain countries to leapfrog many technologies to acquire access (electricity, computing, networking).
- Gender: some uses of the net, e.g. e-mail and chat lines encourage girls to get more involved.
- Political: because of the free flow of information some countries are forced to become more open.
- Cultural: more 'broadcasters' and broadcast opportunities are available so more cultures can speak for themselves and explain themselves; especially when the bandwidth is available, multimedia applications open opportunities for more varied cultural expressions.
- Educational: the Net opens access to educational texts and information about learning that are up to date and freely downloadable. Users of the Internet can also freely join global discussion lists.

In view of the above the following recommendations are made:

- The technology should be expanded so that multimedia can be used to overcome linguistic barriers.

- Educational systems in the West should open up and learn about cultures of the world.
- There is a need to help poorer countries to access ICT using alternative technologies - satellite phones, solar-powered laptops and cheaper computers.
- There is a need to create a culture of inclusion for other cultures and females.
- Developing countries must be included to participate in global trade markets.

Acknowledgement

The above is based on the report of the SECIII Working Group on Equity and the digital divide. Participants in this Working Group were: Johannes Magenheim (Chair), Alnaaz Kassam (Rapporteur), and others.

6. IMPORTED CULTURE

Imported culture can be defined as cultural imperialism where dominant cultures take over less dominant ones. The following facts can be stated:

- Dominant cultures, for example the Coca Cola culture, do not necessarily reflect the true American identity. It is actually very commercial and stereotypical 'culture'.
- American identity is diverse and includes many cultures.
- The Internet has the power to include everyone if it is used pro-actively.
- Strong commercial influences may drown all diversity.
- Various diasporas (immigrants) could use the Internet to develop their own cultures in their new countries and could introduce new ideas into their home countries.

It raises a number of questions:

- How important is it to preserve individual cultures?
- How important is it to have diversity?
- There is no one right way.
- Have civilisations always enriched each other?
- Is it important to preserve linguistic, cultural and religious diversity?
- Can we use the Net to revive and preserve cultures and languages?
- Are languages and cultures equal in importance?

The following recommendations can be made:

- It is important to re-define American culture so that it is more diverse and less commercialised.
- It is vital that we do not lose languages because they are repositories of culture.

- It is important to include minorities, females and special needs in the decision making process in powerful media circles so as to be more inclusive.
- It is important to develop educational materials which include global diversity and include a range of authors.
- Other cultures should be encouraged to tell their own stories and use/add to the Internet so that it exhibits a rich content of oral cultures.
- Create search engines that discriminate less and are more inclusive.
- In these early days of networked technologies, it is vital that we include all members of society and that the flow of information is from the less dominant to more dominant people.
- There is a real danger that the commercialism of traditional media will take over the new media; we must be vigilant and ensure this does not happen.

Acknowledgement

The above is based on the report of the SECIII Working Group on Imported culture. Participants included Carolyn Dowling (Chair) and Alnaaz Kassam (Rapporteur).

7. CYBER RISK, SECURITY AND PRIVACY

The following observations can be made:

- There are surveillance cameras everywhere.
- Governments are checking e-mails.
- Encryption of e-mails has been possible for more than 15 years but is not applied.
- Manipulating private data or stolen pieces of new technologies is becoming a problem.
- There are hidden links between mobile industries, factories and network providers.
- There are issues related to websites and privacy and updated and dynamic information is difficult to control in real time.
- People who create systems without risks never communicate their opinions with people who are aware of the risks (identity card, mobile telephone, controlling movement through GPS).
- Everybody feels suspicious about ICT, but only a few do something about it (copyright, giving your bank account).
- It is sometimes dangerous to rely on technology and often digital technology and human factors don't mix very well - e.g. the Konstanz/Bodensee aircrash.

Society should be asking a variety of questions, including:

- Should technology be used before public concerns or objections have been assuaged e.g. E-voting, identity cards?
- Should systems, which preserve privacy, that have developed over many years, even many centuries, be swept aside by technology?
- If a technology makes something possible, can human rights be bypassed e.g. digital pictures placed without permission on the web, doctoring of images?
- Should we be taking responsibility or risks?
- How should we educate the users?
- Is informational self-determination a part of human rights?

The following recommendations can be made:

- We do not have to create a system we cannot control (the role of agents in information systems).
- Open sources are the best guarantees to uncontrolled agents.
- Take care with cookies and customer profiles.
- There should be no advertising of/on and no link to an Internet service without an agreement guaranteeing protection of privacy.
- We have to protect minors in the educational system (there is a major difference between children in school and employees).
- We have to inform and to educate the user.
- We have to educate the user to understand a code of ethics.

Recommended actions:

- Use practical exercises, and simulations of cyber societies (role playing, case studies) to train students on cyber risk (on both an emotional and cognitive level),
- Develop discussion on general issues of ethics, freedom, responsibility and risks.

Acknowledgement

The above is based on the report of the SECIII Working Group on Cyber risk, security and privacy. Participants were: Hélène Godinet (Chair), Johannes Magenheim (Rapporteur) and others.

8. ICT LEARNING ENVIRONMENTS

The following observations can be made:

- ICT learning environments promote effective learning.
- ICT learning environments can be used both in dedicated laboratories, in conventional classrooms and at home.
- Both students and teachers require training in the use of ICT learning environments so they can be used to full advantage.

- More social teacher/student interaction is needed for effective learning.
- Teachers and students should spend more time together on tasks/activities.

Recommendations on the basis of these observations are:

- Integrate ICT learning environments into course design.
- Design learning activities for use in laboratories and at home.
- Provide close articulation of course materials.
- Learning activities must be effectively guided.
- Provide introductions and constant follow-up, as well as new materials.
- Ensure the focus is more on learning than on visual materials and the technology itself.
- Prepare learning materials as an enhancement, not as a replacement.
- Be prepared for questions that will be asked by the learners.
- Promote effective time management among teachers and students.
- Reserve extra time for ICT learning environment related tasks and activities.

Acknowledgement

The above is based on the report of the SECIII Working Group on ICT learning environments. Participants were: Hélène Godinet (Chair), Guido Rößling (Rapporteur), and Christino Carbonell.

9. E-LEARNING

E-learning is not just delivery of knowledge to students in a cost-effective way - it is learning in full awareness of an ICT-rich environment. The key question is: Why should learning change? There are several reasons:

- E-learning is more effective.
- We have a new world order and a society, which has changed markedly over the past twenty years. A new world and a new society demands new approaches to learning and E-learning is very much in tune with our e-society.
- World wide demand for education is greater than we can satisfy with the teachers we can provide, especially in developing countries.
- Students are now learning part-time (often because of financial necessity).
- Society is convinced of the need for lifelong learning and knowledge can be out of date in 2 - 3 years.
- We need the availability of e-learning.

We should make use of the knowledge students gain from the work they do while they are learning. Learning should be in teams some of the time because learning is a social activity. However, we must also remember that learning is still a very personal activity.

How should learning change? ICT offers flexibility, empowerment, a change in the style of thinking and an attractive, rich environment, especially for young people. Teachers do not seem to have grasped this. Too many teachers equate E-learning with the web. They direct pupils to look for web sites and provide no support in searching, evaluating, structuring the approach, or the information the pupils uncover. We need to change the way we handle teachers and learners. We also need to capture the imagination and support of parents to back any changes which are made (see David Wood's Keynote paper). Despite considerable expenditure the infrastructure in schools is still a problem.

Acknowledgement

The above is based on the report of the SECIII Working Group on E-learning. Participants were: Yvonne Büttner (Chair), Barry Blakeley (Rapporteur), Beat Döbeli, Dirk Draheim, Michael Janneck, Peter Micheuz, Robert Munro, Prathiba Nagabhushan, Clara Oliveira, Ulrik Schroeder, Alnaaz Kassam, Tom van Weert.

Intelligent agents in an e-literate society: some ethical considerations

Carolyn Dowling
Australian Catholic University, 115 Victoria Parade, Fitzroy 3065, Victoria, Australia

c.dowling@patrick.acu.edu.au

Key words: Ethical/Ethics, Future, Internet, Social Issues, Values

Abstract: One of the fastest growing applications of AI research is the implementation of computer programs commonly referred to as 'agents'. This type of software is distinguished from more traditional programs by a high degree of autonomy in decision making and action, the ability to 'learn' from experience and to adapt their behaviour accordingly, and often a highly personified interface. Many are specifically designed to process complex information, make decisions and initiate actions in 'mission critical' areas of human endeavour including health, scientific research, government, business, defence, the law and increasingly in education. While in some cases we are aware of our interactions with these electronic entities, in many contexts their activity takes place 'behind the scenes', at a level not apparent to the user. Implicit in our conception of an agent both in the physical world and in cyberspace is the notion of delegation. Important aspects of this concept are our understandings of features of human interaction such as trust, responsibility, privacy and our capacity to judge competence and intention. Consideration of these issues in relation to the activities of software agents could lead to the formulation of a broadly based code of 'agent ethics'. This could help in regulating some aspects of agent behaviour and act as a foundation upon which common expectations on the part of users might be formulated.

1. INTRODUCTION

Until relatively recently, the notion of intelligent software agents belonged in the realms of speculation and esoteric research. Today it is one of the fastest growing applications of Artificial Intelligence and is already a well accepted aspect of our everyday computing experiences. Some agents, such as the variety of 'personalities' available to assist us with anything from the smallest to the most complex computer based task, are visible to us and are, to a certain extent, amenable to our direct control. Others, particularly in online environments, undertake their activities at a level not apparent to the casual user.

These programs have several distinguishing features in common. One of the most important is a high degree of autonomy or independence. In computing contexts this means that the program is able to make decisions and initiate actions without the need for constant monitoring and intervention on the part of a human being. Another key attribute is the ability to 'learn' from experience and to respond flexibly to changing situations. This may cause such programs to behave in ways not clearly predicted by the original programmers. For many purposes, a highly personified interface suggesting the possession of a believable 'character' or personality provides an important basis for the agent's interaction with users.

Early definitions of agent software, based more on speculation than on experience, tended to emphasise the broader 'human' aspects of these programs, for example:

"an agent is a character, enacted by the computer, who acts on behalf of the user in a virtual environment" considered useful in mediating "a relationship between the labyrinthine precision of computers and the fuzzy complexity of man." (Laurel 1990)

Typical of more functional definitions is:

"An agent can be viewed as an object which has a goal and autonomously solves problems through interaction, such as collaboration, competition, negotiation and so on." (Kinoshita and Sugawara 1995)

From another perspective, an agent exhibiting these characteristics can be described as having:

"a set of beliefs about its environment and about itself; a set of desires which are computational states which it wants to maintain, and a set of intentions which are computational states which the agent is trying to achieve." (O'Hare 2000)

The widespread implementation of agent based systems raises a number of issues that have clear ethical, even moral dimensions. They include the balance between autonomy and control and questions related to trust, responsibility and privacy.

2. IMPLICATIONS OF THE 'AGENT' METAPHOR

Why have we been so quick to accept software agents as an essentially unproblematic element in our computing environment? A large number of our day to day activities currently take place within computing environments characterised by rapid change, large quantities of extraordinarily complex information and a lack of common organisational structures through which information may be accessed and managed. It is clear that, as Laurel predicted, there are now many situations in which some form of 'intelligent' mediation is required between computer systems and the needs of users. Agents are a means of masking the complexity that lies behind our use of computers, thereby facilitating the achievement of our various aims and goals.

The term 'agent' strikes a comforting note of familiarity with many computer users, most commonly conjuring up an image of a docile and amenable servant undertaking particular tasks at the behest and in the interests of the user. As a way of conceptualising our relationship with the technology it has a great deal in common with an earlier metaphor which proved extremely popular, that of the computer as a 'tool'. Both of these images encompass the reasssuring suggestion that the technology is under the control of the user. In each case, however, the metaphor is less straightforward than it appears and the confidence and security engendered in users may well be misplaced. In many instances the notion of an agent masks not just a functional complexity, but also a range of activities which are not of the users' choosing and not necessarily even in their interest.

Both in the physical world and in cyberspace, the concept of an agent is understood and instantiated in a number of different ways. Some of the most interesting types of human 'agents' to consider in relation to the ethical and moral complexities of the roles undertaken by software agents are those which actually represent the interests of more than one party and must effect a balance between competing claims. Examples in the physical world include estate agents, theatrical or literary agents and employment agents.

Another type of agent with which we are all familiar but are probably less comfortable, is the 'secret agent', generally a gatherer of supposedly confidential information operating under an assumed identity. This model is particularly relevant to some current practices on the Internet. Far from being initiated through a contractual arrangement entered into voluntarily by the user, interaction with many such agents is an invisible and frequently unsought component of an apparently innocuous task such as a search for information.

Common to all of the understandings of the roles and functions of human agents outlined above is the assumption that their usefulness derives from the possession of specialised skills, which qualify them to mediate between an individual and a particular environment for the more effective achievement of various ends. Furthermore, there exists a commonly held understanding that most human agents can be relied upon to act in accordance with an understood set of ethical principles that ensure the client or user is not disadvantaged. Such assumptions play a significant part in encouraging the acceptance of programs characterised as 'agents'.

3. THE ROLE OF PERSONIFICATION

In circumstances where the user is intended to be aware of the activities of the software and to interact knowingly with it, the agent metaphor lends itself particularly well to reinforcement through a personified interface. Anthropomorphic elements have been implicit in most computer interfaces from the earliest days. Intelligence and language use are widely accepted as the key criteria distinguishing us both from other living things and from inanimate objects. The very use of the term 'intelligence', albeit 'artificial', in relation to computer programs, in combination with the fact that for the most part we interact with them through language, adds support to the perception that a software agent is 'one of us', and should be subject to the types of expectations, including that of trust, that govern our interactions with our fellow human beings.

Not only are we accustomed to interacting with computers as though they share with us a degree of 'humanity', but in a number of areas of activity we value 'social interaction' particularly highly. An example is the field of education, where our current understandings of learning depend very much on an acceptance of the role of the social construction of knowledge. The high value we place on the 'social' also lends acceptability to the interlocking activities of multiple agents when they are metaphorically characterised as a 'society' (Franklin and Graesser 1996; Costa and Perkusich 1997).

However, while we might acknowledge as inevitable the inadvertent attribution of human qualities to computers and their programs, the deliberate cultivation of personified or anthropomorphic interfaces is more problematic. Both research and experience suggest that a mismatch between realism in appearance and the apparent knowledge level of the agent as revealed through its use of language and other capabilities can have a deleterious effect on credibility and on acceptance. As Masterton, writes:

"A common problem with AI programs that interact with humans is that they must present themselves in a way that reflects their ability. Where there is a conflict between the ability of the system and the users' perception of that ability a breakdown occurs and users may either fail to exploit its full potential or become frustrated with its shortcomings." (Masterton 1998)

Agents that 'look' smart and 'act' or 'talk' dumb are poorly received by many users, who express a higher tolerance for the limitations of a 'character' more sketchily represented, for instance through cartoon-like graphics, including those characterised as normally inanimate objects or as animals (Chan 1998). This may provide some counterbalance to our tendency to unquestioningly invest a high level of trust in such entities. Here too, however, there are complexities, although of a psychological rather than an ethical nature. While in many instances we may feel more comfortable giving instructions to a 'character' that is not fully human, we may not feel the same way about receiving advice or even correction from such an entity. Users of Microsoft Office will be all too familiar with the indignity of being subjected to the whims of an animated paperclip.

The issue of the degree to which software agents can and should be personified is extremely important when considering the ethical aspects of agent systems. It is this aspect of agency that propels questions of right and wrong to a level beyond that pertaining to our use of more traditional types of programs. To the extent that personification is successfully implemented, the expectations of users in relation to the behaviour of the program move into the arena governing the understandings upon which we base our interpersonal interactions, rather than merely those upon which we base our use of technology. The problems are accentuated in cases where the agent is not depicted visually but emulates human dialogue to the extent that users may not be aware that they are interacting with a computer program. Such a situation is inherently and deliberately misleading and is a poor basis for the development of the type of trust in agency that arguably should be an important part of a fruitful ongoing relationship between human beings and software agents.

4. AUTONOMY, TRUST AND RESPONSIBILITY

The delegation of any task to a software agent raises questions in relation to its autonomy of action and decision, the degree of trust which can be vested in its outcomes and the location of responsibility, both moral and legal, for those outcomes.

Many of these agents are specifically designed to weigh up complex information, make decisions and initiate actions in 'mission critical' areas of human endeavour including government, health, scientific research, commerce, the law and defence, with a significant degree of autonomy being intrinsic to their usefulness. Implicit in the recognition of the need for a capacity to exercise initiative is an acknowledgment that some outcomes of agent activity may not be easily predictable by the user. In some cases they may even be contrary to what the user might perceive as his/her interests and wishes.

A further issue emanating from consideration of agent autonomy is that of responsibility for outcomes and actions resulting from decisions which are out of the user's control. Indeed they may relate to capacities of the program of which the user had no knowledge at all. It is inevitable that such instances will occur, precipitating the need for renewed examination both of community and legal understandings of liability.

Closely associated with our ambivalence towards agent autonomy is the issue of trust. How might we decide whether or not to 'trust' an agent? In the physical world a reputation for credibility in most fields is contingent on a verifiable history demonstrating qualities such as accuracy, reliability and efficiency. Where such assurances are available in relation to the software agents to whom we delegate responsibility, perhaps we can be justified in taking the risk of 'trusting' them. As with some forms of delegation to other human beings, however, we may have to accept that there are no absolute guarantees.

Even if we are prepared to trust the agents with whom we deal directly, a further question remains as to how agents might reasonably 'decide' how to trust one another. Implications for users include important considerations in regard to privacy. For instance, the main threats to privacy have been defined by one researcher as:

"Threats caused by agents acting on behalf of the user (through loss of control of the user on (sic) his agent and the disclosure of the user's personal information)" and "Threats caused by foreign agents that act on behalf of others (information extraction via traffic flow monitoring, data mining and even covert attempts to obtain personal information directly from the user's agent)." (Van der Lubbe 2000)

5. A CODE OF ETHICS FOR SOFTWARE AGENTS?

An important element in our decisions to invest trust in expert human agents is our belief that their actions will be governed by a code or codes of ethical behaviour, both at a personal and a professional level. Adherence to a code of ethics is, in fact, a key element in the definition of a profession.

What might a code of ethics for software agents look like? Given that they could well be described as the disembodied 'robots' of cyberspace, an appropriate place to start might be Isaac Asimov's Three Laws of Robotics, familiar to several generations of science fiction readers. A generalised version of the original laws is the directive that a robot may not injure humanity or, through inaction, allow humanity to come to harm.

AI researcher Marvin Minsky acknowledges issues similar to those recognised by Asimov in relation specifically to software agents when he writes:

"There's the old paradox of having a very smart slave. If you keep the slave from learning too much, you are limiting its usefulness. But, if you help it to become smarter than you are, then you may not be able to trust it not to make better plans for itself than it does for you." (Minsky1994)

Some current research in this area expresses the problem in terms of the need for the types of 'norms' that constrain behaviour within human society (Verhagen 2000).

6. CONCLUSION

There is no doubt that the use of those AI applications we refer to as 'agents' is increasing in a range of areas critical to our well being both as individuals and as members of society. Without necessarily implying any intent on the part of software developers to promote agent behaviours which we would regard as unethical, it would seem prudent for ethicists and programmers to collaboratively give serious consideration to the formulation of an appropriate ethical code that, if accepted, might give users of agent technologies a level of confidence in the performance of these entities comparable with that which is currently vested in human beings.

REFERENCES

Chan, T. -W. (1998), *The past, present, and future of educational agents.* [http://www.apc.src.ncu.edu.tw/apc/ppt_chan.html]

Costa, E.deB. and Perkusich, A. (1997) Designing a Multi-Agent Interactive Learning Environment. In *Proceedings of ICCE 1997*, International Conference on Computers in Education, Z. Halim, T. Ottmann and Z. Razak (eds.), Charlottesville, VA.

Franklin, S. & Graesser, A. (1996) Is it an Agent, or just a Program?: A Taxonomy for Autonomous Agents. In *Proceedings of the Third International Workshop on Agent Theories, Architectures and Languages*, Springer-Verlag, Berlin.

Kinoshita, T. and Sugawara, K. (1995) *Agent oriented computing*. Soft Research Centre, Tokyo.

Laurel, B. (1990) Interface agents: metaphors with character. In *The Art of Human-Computer Interface Design*, B. Laurel (ed.), Addison-Wesley, Reading, Massachusetts.

Masterton, S. (1998) Computer support for learners using intelligent educational agents: the way forward. In *Global Education on the Net Vol. 1*, T. W. Chan, A. Collins and J. Lin (eds.), Proceedings of ICCE '98, the Sixth International Conference on Computers in Education. China Higher Education Press and Springer-Verlag.

Minsky, M. (1994) A conversation with Marvin Minsky about Agents, *Communications of the ACM*, July 1994, vol. 37, no. 7.

O'Hare, G. (2000) Agency, Mobility and Virtuality: A Necessary Synergy. In *Proceedings of ISA 2000*, International ICSC Congress on Intelligent Systems and Applications. ICSC Academic Press, Canada, pp. 15 -21.

van der Lubbe, J. (2000) Incorporating Privacy Enhancing Technology in Intelligent Software Agents. In *Proceedings of ISA 2000*, International ICSC Congress on Intelligent Systems and Applications. ICSC Academic Press, Canada, pp. 222-228.

Verhagen, H. (2000) Trust and Norms for Artificial Agents. . In *Proceedings of ISA 2000*, International ICSC Congress on Intelligent Systems and Applications. ICSC Academic Press, Canada, pp. 277-280.

Critical thinking and an ethical approach to studying History - the case for ICT

Alnazz Kassam
Toronto District School Board, 8 Burnt Bark Drive, Scarborough, Canada M1 V3 J8

alnkassam@yahoo.com, akassam@oise.utoronto.ca

Key words: Culture, ICT, Justice, Society

Abstract: In this paper it is argued that the introduction of the Internet into the classroom provides a unique forum for the cultures of the world to communicate with each other, arrive at a new understanding between each other and be able to re-define views of how knowledge is formed. The paper will present two alternate views of critical thinking - the orthodox and the second cycle - and their connection to historical inquiry. The paper will show how these two approaches to critical thinking are reflected in an Internet project, which allowed eight Canadian classrooms to study history from a First Nations (indigenous peoples) perspective.

1. INTRODUCTION

The research data for this paper stemmed from the Other Story project (funded by the Ontario Ministry of Education from 1993 to 1995). This was designed to bring the perspective of the "other" into the classroom. The project argued that in textbooks and through the media, cultures of the world are often reflected in ways that their members might not agree with.

In this paper I will show how the project, in offering an opportunity to explore and reflect, helped re-define our approach to education, to the study of history and to critical thinking.

2. CRITICAL THINKING AS GOOD THINKING

Critical thinking, in recent years, has acquired popularity as a way of getting students to be independent thinkers able to function in a democracy.

"Most orthodox theoretical accounts of critical thinking argue that the ultimate function of good thinking is to distinguish between justified and unjustified claims of beliefs. This is done by applying the rules and techniques of formal and informal logic to propositional expressions in order to determine if their statements are true and their arguments valid or sound. Justified claims and beliefs, then, reduce to those that stand up to the rigorous tests of logical analysis, while unjustified ones, obviously do not. A good thinker, consequently, is one who is skilled in the manipulation of logical criteria and who willingly abides by the evaluative conclusions it generates." (Walters 1994)

Giroux, however, considered two major assumptions were missing from orthodox definitions of critical thinking - the relationship between theory and facts and that knowledge cannot be separated from human interests, norms and values.

"How information is selected, arranged, and sequenced to construct a picture of contemporary or historical reality is more than a cognitive operation; it is also a process intimately connected to the beliefs and values that guide one's life." (Giroux 1994)

Obviously historical analysis cannot be dismissed as value-free or as a simple explanation of "what happened". In the choice of facts and the framework provided in their interpretation we make decisions about how we interpret the world and our place in it as players.

3. PROJECT OVERVIEW AND ACTIVITY

The Other Story was a research project to develop curriculum for inter-cultural understanding through a critical inquiry, issues-oriented approach. It combined classroom-based discussion and an on-line conference linking schools throughout Canada to students, teachers and adherents of the culture being studied. The project's most important objective was to study cultures from their own perspective. It was envisaged that through on-line activity the students would develop a deeper, more sophisticated and less media-bound understanding of different cultures.

In Spring 1995, eight Grade 8 classrooms, including two from First Nations schools, participated in an Internet conference "First Nations Peoples: The Untold Story". This examined a significant event in Canadian history and presented it to students both from the conventional textbook view and from the perspective of the First Nations people. Students read and responded to a story written by two First Nations authors about Mistahimaskwa (Big Bear) who defied the encroaching Canadian government forces who wished to occupy and take over his people's land. Students were encouraged to look for information in other resources and consider: Have we been exposed to the First Nations viewpoint before? Why or why not? What does this tell us about our history and about ourselves?

Responses from each classroom were shared on a listserv. One project objective was to encourage reflection - have we as teachers, students and adherents of the various cultures learned anything from this exchange of experiences? How does this portrayal of indigenous and non-indigenous peoples affect our relationship? How can we use this learning experience to affect a change in the relationship between us?

This paper focuses on the student responses from three very different schools - A. S. Dawson, Shoreside and Lakeside.

3.1 A. S. Dawson's Participation in the project

This is a suburban school in a large metropolitan city. It stands in the middle of a first generation immigrant neighbourhood close to a wealthy residential area. The students are hard-working and well above their grade level in reading and computational skills. They introduced themselves clearly confident of their strengths and their ability to see the "other" view:

"As students, we have been influenced in many ways. We watch videos and take trips to educational areas. These help expand our thoughts. By having these privileges, it helps us think more clearly and assists us in deciphering what's right and wrong. This enables us to see the Native's aspect of the situation."

3.2 A. S. Dawson and Mistahimaskwa's story

The story of Mistahimaskwa, is the story of a Cree chief, who, during the 1870's, resisted treaty agreements with the Canadian government and attempted to form a united front against government appropriation of aboriginal land (Fletcher 1995). Mistahimaskwa does not succeed in keeping his land. In fact his attempts at compromise and his treaty arrangements with the government result in betrayal. This story is very much a reflection of Canada's history - who we are and where we have come from.

The story was delivered to the students on-line and their first response was:

"When we first read the story, we had very strong and intense feelings about it. We felt very sympathetic towards the Cree and their chief, and we felt very outraged towards the Europeans who tried to take away the land which the Cree lived on."

The students noted the wanton killing of the buffalo by the Europeans, the "heartless and ignorant" snatching away of the Cree land, the fact that the Europeans didn't uphold the treaty and the lack of understanding they betrayed toward the Cree culture:

"If the Europeans had tried to put themselves in the Cree position, history probably would have changed for the better. We feel the Europeans acted unfairly towards the Cree, treating them as if they were inferior."

Initial analysis of the language used in this passage displays two examples of bias. Firstly, students do not believe the perpetrators of this injustice are the Canadian government. They identify them as Europeans. From this story and their study of history, the students should be well aware that the Canadian government sanctioned land acquisition. The Canadian government dispatched 8000 troops to vanquish Mistahimaskwa's several hundred tired, hungry men. The students didn't notice this. Distinguishing between European and Canadian is an attempt to distance the actions of the perpetrators from the students' own government and thus from themselves (Fletcher 1995).

The second example of bias is the underlying tone that the Europeans could not be "blamed" for what happened because they did not "understand" the other culture - if they had not perceived the Cree to be inferior, they would not have treated them in this manner. The greed and injustice of the situation is downplayed - a refusal to acknowledge responsibility and guilt on the part of one's government.

After examining other resources the students commented:

"We compared the story from the text book and the story by Dion and Fletcher. Some similarities mentioned in both stories included the way the native Indians were acknowledged as poor and starving from the lack of buffalo. Surprisingly, the textbooks also looked at the grievances the natives had. The location of the battles specified in each story was also identical. Both mentioned that Big Bear was tried and found guilty of the deaths caused by his warriors. The similarities the stories shared were basically the clean cut facts."

The most important conclusion the students reached was the determination of which text was more accurate. This satisfies the first approach to critical thinking - examination of the logic of the arguments and the verity of the facts presented by both parties. The students identified the similarities between the texts as being the clear-cut facts. Facts can never stand alone - in choosing the facts to include in their version of events historians from both cultural groups made choices. The facts, rather than being neutral and value-free, become in and of themselves an extension of the historians' perspective and interpretation (Carr 1972). For these students, factual analysis is important - the emotional response of the Cree is not recognised for what it is - the unfairness of the situation.

Susan Fletcher was interested in the students' interpretation of the event - in what they thought it told them about themselves as human beings, as Canadians. She felt they couldn't complete this process until they examined the interpretation of the event from both perspectives - "those who did it and who it was done to". The students, in avoiding the question of what happened in the larger context of whether it was right that it should have happened also avoided bringing themselves and their own positions into the story. In understanding history from the first critical thinking perspective of logic, they avoid the larger question of whether the government acted fairly and the question of how the historians interpreted this event.

3.3 Shoreside's students and Mistahimaskwa's story

These students interpreted the same story in a totally different way - which led to different insights. After comparing the story with what they found in their books they commented:

" 'the Cree killed nine whites at Frog Lake, burned Ft. Pitt, and were defeated at Loon Lake'. The article in the encyclopaedia does not include the fact that eight thousand government troops were sent after them. We think that one reason this was not mentioned was because the author of the article did not want to make the Canadian government look bad."

Here students are looking at the historians' "intention" and bias. Only some facts are reported so that the Canadian government is not portrayed as evil. This is a form of critical thinking that goes beyond consideration of the given facts. The students then go beyond motive for interpretation of how the vocabulary and images one uses affect how one views events and people.

"In the encyclopaedia "Encarta '95", a brief article on the Cree Indians states that they practised cannibalism in times of hardship. Cannibalism is one of the quick reference words that shows up 'highlighted'!"

The students are beginning to understand the use of vocabulary to represent reality. They realise that linking "cannibalism", an entirely unjustified connection, will affect the way that the Cree are perceived. As they indulge in criticism and in historical analysis of this order they are becoming historians in their own right - no longer accepting history as mere physical events but rather as explanations and interpretations of past events by the historians of their text books. Their intellectual development is illustrated by this response:

"We now see language has a big impact on us and different words, even if they mean the same thing, can give a whole new meaning to a statement. We now know not to believe just one account. Depending on who wrote the account seems to colour what facts are recorded or not recorded."

I would argue that by juxtaposing the text of the textbook, the encyclopaedia and the field visit with the text of the "other", fresh vistas of criticism are revealed. Students begin to question why they have never seen this perspective before.

"Recently our class went to Fort York and stayed overnight in one of the blockhouses. When we looked through the books we learned the views of the soldiers, the officers and Mrs. Simcoe. At no time did we read about or hear from the Fort York interpretative staff what the native point of view was. It's like this in many historic accounts in our books. We were wondering if there was a website or e-mail address where we could get more information."

These students are beginning to move away from the textbooks alone to what their own society provides as the official history of the relationship between the two peoples. They're beginning to see ICT as a possible source for information. They are also beginning to question formal knowledge - the knowledge of their own society and whether it portrays what really happened.

3.4 Lakeside School and the Story of Mistahimaskwa

When we meet the Lakeside students we begin to understand that critical thinking cannot be part of history until one understands the impact of history on a people (especially people without power and subject to the whim and will of a sovereign government). This is a First Nations school in a small town in Northern Manitoba. The students introduced themselves:

"We are all in grade 8. There are 23 students: 11 boys and 12 girls. We live in Lakeside, Manitoba. It has about 1000 people, three stores, one cafe, and one small motel. Languages spoken are Cree and English. In our recent past (mid 1970s) Manitoba Hydro flooded the lake and river system we live on. Our entire community was moved to higher ground. Our old homes were burned and the fishing industry was devastated. Our people are still recovering from the disaster."

They identified closely with the story of Mistahimaskwa.

"The story makes me feel sorry and angry at the same time. Sorry for the people whose lives were being destroyed and changing the way they've lived for years, because they really had no choice. Angry because those people whose lives were being destroyed are a part of me and I am a part of them. Today things like that are still happening. It was sad because they tried to accept the government's promises but they were not honoured until they lost their temper and killed a lot of people. The government over reacted and sent 8000 people to capture 200 of Mistahimaskwa's people."

Two powerful issues emerge from this interpretation of the story. Firstly, it is their story. They cannot help but identify with the feeling of fear, hunger and loss of one's land, culture and heritage - issues of objectivity and impartiality are secondary. The second related issue is that the oppression is continuing to happen to them today. Participation of the Cree students should bring alive to non-First Nations students the political importance of the story of Mistahimaskwa. This political perspective is present in all historical analysis of events that involve peoples of disparate powers.

The story of Mistahimaskwa cannot remain the historical analysis of an event - an impartial, objective reporting of "what happened". In its interpretation lies the explanation of what Canada was and, through the story of the students at Lakeside, the interpretation of what Canada continues to be today. This interpretation cannot be impartial and objective - like all other history it has a moral dimension. In that context the story as told by the victim becomes critical. Critical thinking is impossible until and unless you juxtapose the view of the dominant group with the view of the "other". The whole story cannot emerge until both stories are told. It is in placing oneself as part of this history, in critically examining one's position and identity within this story that true critical thinking can take place.

4. CONCLUSION

How do we analyse the Lakeside students' participation in the re-creation of this historical event? I believe that in the listening and in the creation of the audience for the stories of the other through ICT, we are providing a new context for understanding of the other and also of ourselves. History is not the history of the dominant group - it is the history of the relationship created by that domination. We cannot understand that domination unless we see it from the perspectives of both dominated and dominator. The Internet provides a unique environment to juxtapose these two realities providing more complexity to one's acquisition of knowledge and one's critical analysis. This juxtaposition - of the story of the "other" with the "official" version - brings deeper understanding of critical thinking. This interpretation of critical thinking goes beyond objectivity but acknowledges that when human beings explain events they make meaning of them. When you have two contradictory interpretations of one event - especially an event that involves injustice - then true critical thinking will involve debate on one's place in the carrying out of this injustice. I believe that ICT provides a unique forum for multiple perspectives and for egalitarian debate.

REFERENCES

Carr, E.H. (1972) The Historian and His Facts. In *Teaching History in Canada*, G. Milburn (ed.), McGraw-Hill Ryerson, Toronto.

Fletcher, S. (1996) *I never realized how much I didn't know, Sharing the Story of Mistahimaskwa with Intermediate Students on the Internet*. Unpublished paper.

Giroux, H. A. (1994) Toward a Pedagogy of Critical thinking. In *Re-thinking Reason: New Perspectives in Critical thinking*, K. S. Walters (ed.), State University of New York Press, New York.

Walters, K. S. (ed.) (1994) *Re-thinking Reason: New Perspectives in Critical Thinking*. State University of New York Press, New York.

A look at the impact of ICT on the informational power relationship between corporations and consumers

Christopher Lueg

Department of Information Systems, University of Technology Sydney, PO Box 123, Broadway NSW 2007, Australia

lueg@it.uts.edu.au

Key words: Agent of Change, Cyber Surveillance, Electronic Communication, Globalisation, Internet

Abstract: Prior to the proliferation of access to global computer networks, people were mainly passive consumers of corporate information disseminated via centrally owned and operated media, such as broadcasting, television, and newspapers. Branding is a marketing technique that has largely benefited from this uni-directional information flow to consumers from corporate information sources. Branding aims at positioning products in the consumer's mind by associating a product, a service or a corporation with a unique name, logo, or selling theme, such as belonging to a certain group because of consuming a particular product. ICT, such as the Internet and the World Wide Web in particular, have enabled ordinary people to disseminate information to a large audience as well as to retrieve information from a variety of information sources beyond corporate spheres of influence. This means that ICT directly impact the informational power relationship between corporations and consumers. This paper examines the current role of information and communication technologies in this context, and speculates on future roles of ICT.

1. INTRODUCTION

The importance of ICT has increased significantly over the past few years. The daily lives of people are increasingly affected by information technology and network-related services, and the rise of the network society is regularly discussed. Compared to the prevailing term "information society", "network society" pays more attention to the fact that processing information has been important for decades while networks have only recently begun to re-define notions of time and (physical) location. Eriksson (1999) argues that we are witnessing the early phases of a societal transformation that will be as dramatic as the first two industrial revolutions that are commonly associated with steam and railways, and electricity and the automobile, respectively. In analogy, networks have effects similar to those of railways whereas non-networked (but nevertheless information-processing) computers have had effects more similar to those of stationary steam engines.

Global computer networks, such as the Internet, have revolutionised the way information is disseminated and accessed. Prior to the advent of the Internet and the World Wide Web, in particular, people were more or less passive consumers of information disseminated via traditional media, such as broadcasting, television and newspapers. The tools of information flow required large investments and it was easy to identify those controlling these media (Spencer 1997). Information provided by these media along with corporate press releases and advertisements were the most significant ways (and often the only ones) people could use to inform themselves about corporations, their products and the quality of the products. In contrast, the Internet has enabled ordinary people to disseminate information to a large audience and to retrieve information provided by a variety of sources. These range from corporate information to sources provided by competitors to those operated by disgruntled consumers or anti-corporate movements.

This paper explores the role of ICT through the example of branding. This marketing technique has largely benefited from the uni-directional information flow to consumers from corporate information sources. Providing access to a variety of independent information sources means ICT directly impact the informational power relationship between corporations and consumers.

2. THE FIGHT FOR THE CONSUMER'S MIND

Branding denotes "the act of differentiating a product or service or corporation with a unique name, logo, positioning, selling theme, etc."

(Upshaw quoted in Kania 2001). Brands represent promises about what consumers can expect from a product, a service or a company. A different perspective on branding is provided by Downes (1999) describing it as:

"Advertising today looks for stronger hooks, and it finds them in association and self-identification. The concept is especially simple: find (or define) a person's conception of self which is pleasing. Mould that conception such that the use of a product or service is essential to that conception. Imprint the idea that in order to be yourself, you need to purchase such-and-such a brand."

Downes outlines how this is used to sell products:

"The product, no longer able to offer satisfaction on its own ground ("a potato chip is a chip is a chip"), instead offers the consumer a chance to be part of a certain 'crowd' or 'scene'."

Memes play a central role in the fight for the consumer's mind. A meme is a:

"contagious idea that replicates like a virus, passed on from mind to mind. Memes function the same way genes and viruses do, propagating through communication networks and face-to-face contact between people." (Bennahum quoted in Downes 1999).

Others see corporate ideas and slogans as similar to memes that can be used to "influence certain stakeholders' set of beliefs in such a way that they internalise a positive image of the company". (Porak et al 2000). Media play a double-edged role. They can be used to propagate "corporate memes" and for meme-based "counterattacks" - for example as a tool to attack the information flow from corporate media:

"The methodology of counterattack involves inserting counter-memes into the media mainstream. It is the idea of the meme conceived as virus taken to its logical extreme." (Downes 1999).

The power of meme-like activities was demonstrated recently when an MIT student managed to associate the terms "Nike" and "sweatshop". His e-mail exchange with Nike when trying to order running shoes customised with the word "sweatshop" (Peretti 2001) was widely circulated on the Internet and was reported in a range of more traditional media (MediaGuardian 2001). The power of memes is important as is seems that corporate branding strategies do not only provide certain benefits, but also may introduce certain vulnerabilities. Klein (2000) argues that brand images are the corporate Achilles' heel:

"multinationals like Nike, Microsoft and Starbucks have sought to become the chief communicators of all that is good and cherished in our culture: art, sports, community, connection, equality."

Klein argues that the focus on brands has made these corporations the more vulnerable to attacks the more successful they are:

"if brands are indeed intimately entangled with our culture and our identities, when they do wrong their crimes are dismissed as merely the misdemeanours of another corporation trying to make a buck. Instead, many of the people who inhabit their branded worlds feel complicit in their wrong, both guilty and connected. But this connection is a volatile one: it is not the old-style loyalty between lifelong employee and corporate boss; rather, it is a connection more akin to the relationship of fan and celebrity: emotionally intense but shallow enough to turn on a dime."

The Economist (2001) argues that:

"the more companies promote the value of their brands, the more they will need to seem ethically robust and environmentally pure. .... The fact remains that brands give [protesters] far more power over companies than they would have otherwise."

3. BRANDING AND ICT'S DOUBLE-EDGED ROLE

From a corporate perspective, ICT and the World Wide Web in particular have enabled novel ways to address customers and shareholders:

"traditionally, branding messages have been unidirectional, delivered to customers via radio, TV, magazines, and other media. In contrast, the Web allows two-way communication - a dialogue in which buyer and seller learn more about one another." (Kania 2001)

Aaker and Joachimsthaler (2000) argue that "to understand the Web, an experience-based model such as a theme park or a retail store is a better metaphor than passively received advertising." and that interaction with other people on a corporate Web site has the potential to create "a meaningful social experience involving the brand". According to Kania "the Web has erased the limits defining how customers can experience brands".

Anti-corporate movements see a different role for ICT:

"with increased public access to the Internet in the late 80s and early 90s, meme hacking was given a new life. While their access to mainstream media was still limited, activists could now communicate with each other in rapid, free and uncensored messages. Moreover, the Internet and especially the World Wide Web gave them a means of reaching directly into the mainstream consciousness, bypassing the media altogether."

Klein (2000) quotes Peter Verhille of Entente International outlining:

"one of the major strengths of pressure groups - in fact the levelling factor in their confrontation with powerful companies - is their ability to exploit the instruments of the telecommunication revolution. Their agile use of global tools such as the Internet reduces the advantage that corporate budgets provided."

Disseminating certain information outside "regular" channels has shown to be a strong lever. Klein quotes Simon May, Shell's Internet Manager:

"There has been a shift in the balance of power, activists are no longer entirely dependent on the existing media. Shell learned it the hard way with the Brent Spar, when a lot of information was disseminated outside the regular channels."

Information circulated online can have a strong impact on companies and their reputations. Aaker and Joachimsthaler (2000) note that "customers with a bad experience can reach thousands of people instantly at no cost." Companies clearly need to be aware of potentially threatening information that is circulated online. There are even claims by PR people that companies who fail to monitor Internet traffic may be headed for a public relations disaster (Ulfelder 1997). Various reported incidents illustrate the threat potential, for example the designer Tommy Hilfiger was the victim of a threatening urban legend. The legend goes that Hilfiger appeared on the Oprah Winfrey Show, made racist comments about several groups and was then tossed off the set by Winfrey. In fact, Hilfiger has never appeared on or taped an episode of Winfrey's show. However, the legend spread so rapidly and generated so much controversy among customers and potential customers that the company was forced to respond on the Internet. To fully understand the threat potential it is important to know that the brand "Tommy Hilfiger" draws some of its strengths from its popularity among ghetto kids and hip-hop fans (Klein 2000).

4. LIMITATIONS OF ICT

In order to be able to react to threatening information disseminated online, companies need to know what is circulating on the web. This could be interpreted as a third role of ICT - helping re-establish the traditional information relationship between corporations and consumers. Companies, notably eWatch (www.ewatch.com), CyberAlert (www.cyberalert.com) and IntelliSeek (www.intelliseek.com), offer specific tools that help companies search the Web and other information sources to find out what is being said about a company and its products. However, the strengths and weaknesses of commercially available Internet surveillance tools are largely unknown. Furthermore, even the technically most advanced search technology is subject to a variety of limitations (Lueg 2001).

First of all, only a limited number of electronic communication channels are publicly accessible. Among the open channels are Usenet newsgroups, parts of the World Wide Web and public mailing-lists. In contrast, e-mail is mostly private, many mailing-lists are for closed user groups and many web servers have password-protected areas. Secondly, even in the case of the publicly accessible web, it is simply impossible to monitor all traffic for resource reasons (bandwidth, storage capacity, processing power). In 1998, researchers found that coverage of the web by search engines was severely limited: no single search engine indexed more than about one-third of the 'indexable web'. The web has expanded enormously since then which means that coverage can be expected to be much worse. Many web servers generate web pages on the fly, which means that scanning these pages implies scanning the databases the pages are generated from. Thirdly, even state-of-the-art search and retrieval technology is only good at searching for known terms. When people use special nick-names when talking about products or companies it is hard for search technology to find out what people are talking about. Simple examples observed in online discussions are "Wuergerking" (a German language pun) instead of "Burgerking" or "McDoof" (another language pun) instead of "McDonald's". Fourthly, it is difficult for automatic search tools to find out what people actually do when they are talking about things. Removing discussions from their particular social contexts ('de-situating') may result in inaccurate interpretations.

Finally, even if search technology detects mention of corporations or brands, this information is already disseminated. Web sites can be closed but this process takes some time and often the pages are already scanned by search engines, such as Google (http://www.google.com), that store images of indexed pages. Email cannot be 'removed' once disseminated over mailing lists and stored in personal mailboxes. Usenet articles can be deleted on some news servers but many servers do not honour cancelling requests. Private archives are mostly inaccessible.

Of course, ICT is limited on the side of counterattacks as well. In order to be effective, counterattacks need to reach a sufficient number of receptive minds. Disseminating information has become cheap but targeting information and creating a "critical mass" may still be difficult.

5. FUTURE IMPACT OF ICT

As Spencer (1997) observes, the world has changed and there is no going back to the traditional situation where large-scale information flow was easy to control as it relied on scarce, expensive machines. Kania (2001) argues that users go online to find information and that the rich amount of information available online has enabled "super-empowered users". While this clearly is an opportunity for corporations to offer brand information, customers can be expected to access independent information channels as well. It is widely recognised that the Internet has enabled radically new ways for customers to share information but so far it seems that assessment of these alternate communication channels in the marketing literature is based on experiences with more traditional media. We found surprisingly little discussion of the threat potential of ICT in the context of branding. The fact that not only branding and marketing organisations but also counterattack activists make use of human cognitive characteristics by creating and manipulating memes, indicates that corporate influence is limited as long as uncontrollable communication channels exist.

Considering that the power of memes is known, the relationship of ICT, information dissemination and human memory has received surprisingly little attention. It is well known that advertising may create false product beliefs. The manufacturer of a mouth-wash, for example, created the impression that the mouth-wash would be an effective preventative against cold and had been censured by the Federal Trade Commission (FTC) for misleading consumers. What is less known is that human memory may be influenced. It is possible to use information to make people remember past experiences differently or things they never experienced. Braun and Loftus (1998) have shown that although graphical information seems to have a stronger impact, even textual information can be used to influence memories. This impact is important as experiments suggest that consumers use their memories of products when browsing supermarket shelves. Colour of products seems to be extremely important. Under certain circumstances textual information can be used to change memory of colour. It would be interesting to look at potential impacts of these and similar effects in the context of ICT when corporations use ICT to disseminate brand information and when activists use ICT to create counterattack memes.

Based on the information presented in this paper we expect that the virtually unrestricted (and hardly controllable) information flow in the Internet will support what The Economist (2001) predicts:

"Brands of the future will have to stand not only for product quality and a desirable image. They will also have to signal something wholesome about the company behind the brand".

REFERENCES

Aaker, D. and Joachimsthaler, E. (2000) *Brand leadership*. The Free Press. New York, NY, USA.

Braun, K. and Loftus, E. (1998) Advertising's misinformation effect. *Applied Cognitive Psychology*,12, pp. 569-591

Downes, S. (1999) Hacking memes. *First Monday* 4(10).

The Economist (2001) Special Report Brands: September 8th, pp. 9, 26-28.

Eriksson, E. A. (1999) Information Warfare: Hype or Reality? *The Nonproliferation Review*, Spring/Summer. pp. 57-64.

Kania, D. (2001) *BRANDING.COM*. NTC Business Books (In conjunction with the American Marketing Association). Lincolnwood, IL, USA.

Klein, N. (2000) *No logo*. Flamingo. An Imprint of HarperCollins Publishers, London, UK. Citations from the edition published in Australia in 2001.

Lueg, C. (2001) A distributed cognition approach to integrate security management and business processes. In *Proceedings of the 2nd International Conference on "Working with e-Business"*, Perth, WA, Australia.

MediaGuardian (2001). Jonah Peretti and Nike. [http://www.mediaguardian.co.uk/news/story/0,7541,440022,00.html]

Peretti, J. (2001) Culture jamming, memes, social networks, and the emerging media ecology: the "Nike sweatshop email" as an object-to-think-with. [http://www.media.mit.edu/~peretti/nike/]

Porak, V., Geissler, U. and Einwiller, S. (2000) Corporate media: an approach for corporate community management. In *Proceedings of the 8th European Conference on Information Systems*.

Spencer, H. (1997) Age of uncontrolled information flow. *The Information Society* 13, pp. 163-170. Taylor and Francis.

Ulfelder, S. (1997) Lies, damn lies and the Internet. *Computerworld*.

Exploration of Object-Oriented Models in Informatics Education

Torsten Brinda and Sigrid E. Schubert
Didactics of Informatics, University of Dortmund, D-44221 Dortmund, Germany

{brinda, schubert}@ls12.cs.uni-dortmund.de

Key words: Basic Concepts, Didactics of Informatics, e-learning, Learner-Centred-Learning, Learning Environments

Abstract: The authors introduce the exploration of object-oriented basic concepts as a new approach to object-oriented modelling (OOM). While other subjects have a long tradition in learning by discovery with suitable learning aids (e.g. experimental installations of the scientific subjects), Informatics at first has to define and to develop such learning aids. We therefore analysed software development environments and identified the factors that promote understanding and those that inhibit understanding. In particular we found that "support of the syntactic correctness of a model", "promotion of the logical correctness of a model" and "visualisation of abstract concepts with change of the observation perspective" were factors that promoted understanding. Understanding, however, was hindered or inhibited by factors such as "error avoidance by ban of action" as well as the fact that the fundamental ideas of OOM are presupposed. From these findings, design criteria of the Didactics of Informatics for so-called "exploration modules" which are learning aids for the exploration of object-oriented basic concepts, were derived. These key criteria are the "basic concepts on different abstraction levels" and the "synchronisation, transformation and evaluation of diagrams". Finally the role of exploration modules as a lesson medium is discussed and the application of the exploration by learners is explained. The change between different views of an object-oriented model to ensure its completeness and consistency is stressed.

1. MOTIVATION FOR EXPLORATION

Education has to prepare learners for the changing everyday world. Due to their importance in society, the study of informatics systems (particularly computer networks and distributed systems) and object-oriented modelling (as a systematic method of informatics system construction and analysis), are essential components of informatics education.

In the past small programs were developed for informatics education to help learners derive knowledge about real world applications. This worked well as long as both lesson examples and professional solutions were based on the same principles and methods. Learners followed the same strategies as experts, though with objects of greatly reduced complexity. Meanwhile, the algorithms, data structures and software development methods on which large informatics systems are based have changed. Consequently the transfer from 'programming in the small' (simple examples for algorithms and data structures) to 'programming in the large' (principles of software engineering by means of complex examples) is no longer recommended (Applerath et al 1998).

The activities of learners in informatics lessons are frequently limited to the application of standard software and the exploration of problem solving with the help of a programming language. As educational objectives have changed object-oriented programming languages have become an important focus of study. The modelling language 'Unified Modelling Language (UML)' and related tools for analysis and design are increasingly used in informatics lessons. These tools are, however, considered to be less suitable for informatics modelling in informatics education, because the complexity of their user interfaces ties up a lot of lesson time. This is a symptom of the original objective of software development environments (SDEs) - to support professional software developers in producing or constructing efficient and error-free software products (Figure 1).

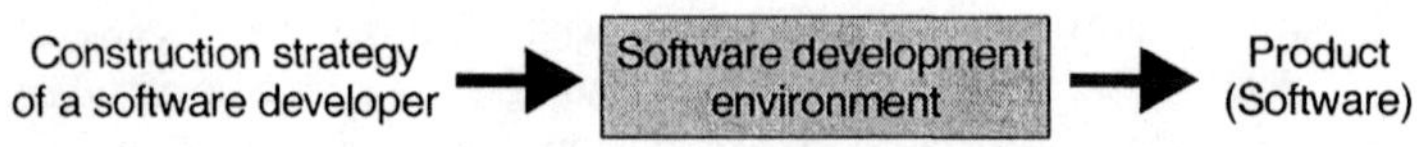

Figure 1. The software development process

SDEs provide functions e.g. 'error avoidance by ban of action' that turn out to be barriers which may be experienced in the educational process of learning. This feature (e.g. selection of the return type of a method from a list of data types) is very useful for the avoidance of certain error classes, but critical if permanently activated from a didactic viewpoint, because then learners are protected from certain error classes and learning by error cannot occur. When working without a SDE or with a SDE that does not offer such functionality, learners become confronted with such errors without having learned corresponding problem solving strategies. In view of the educational process the main problem is that the fundamental ideas of the object of consideration (here: the construction of object-oriented models) are presupposed. These systems address users with pre-knowledge in the design of object-oriented problem solutions and help them to take into account everything essential for modelling, to automate routine tasks and to increase the quality of their solutions. With respect to the teaching and learning of modelling strategies these systems have not been developed, nor do we feel they would be suitable for this task.

Product development and the education of software developers are not objectives of informatics education. Modern concepts for informatics education at secondary and university level stress the active role of the learners to prepare them adequately for self-paced, self-organized and life-long learning. Learning and teaching concepts that should feature more in education are active, collaborative, peer and explorative learning (Turner 1998). Learning aids to support learners in their educational process are essential (Figure 2).

Figure 2. The exploration process

Therefore we developed 'Didactic Systems' as a combination of traditional and modern study concepts. The didactic system for object-oriented modelling makes it possible for learners to navigate in knowledge structures, to construct solutions for exercises from 'Exercise Classes' and to learn by discovery with exploration modules (EMs) (Brinda and Schubert 2001). All these learning forms are embedded in the social processes of the informatics lesson.

2. EXPLORATION WITH SOFTWARE MODULES

The exploration process is extremely important and it is valuable to consider the learning aids that it would be desirable to develop for Informatics. The scientific subjects make extensive use of experimental installations and technology education has a long tradition of working with construction kits.

The Didactics of Informatics group at the University of Dortmund developed a software environment in which learners can construct a part of the Internet architecture in a closed environment and then check the data communication between network components using a protocol simulation (Steinkamp 1999). Learners work directly with a 'construction sheet' window in which they construct and connect components. The data input and output of each component is shown on another window - the 'communication monitor'.

Learners can also see if data communication was not possible due to incorrect connection of components. Learners easily recognize whether or not their construction is correct because the requested web document, or a message to say that this document is not available, appears in a simulated browser window. Through use of this learning aid, it is expected that understanding of the working principles of distributed informatics systems is improved. We intend to develop EMs which we will use in teaching about object-oriented modelling.

3. UNDERSTANDING PROMOTING FUNCTIONS OF SOFTWARE DEVELOPMENT ENVIRIONMENTS

Through the analysis of SDEs, especially modelling tools, the following functions which promote understanding have been identified. These functions are essential for the design of exploration modules.

3.1 Support of the syntactic correctness of a model

Modelling tools provide correct or validated model components as predefined building blocks. The user selects from these correct or validated alternatives. The syntactic correctness of the model remains unchanged as long as a valid combination is chosen. Source code generators are attached to many modelling tools with the related compilers. Successful compilation requires a syntactically correct model. Reports from the system help the learners to identify any syntactic weaknesses in their models. The understanding process is based on, and develops from, the discussion processes within the group of learners.

3.2 Promotion of the logical correctness of a model

Every modelling language is based on a grammar, which defines how the single building blocks may be connected with each other. For example, in a class diagram e.g. inheritance cycles are not permitted. The user is informed when construction rules are disregarded. Typical classes of logical errors are avoided. The assurance of the completeness and consistency of models also belongs to this field. Certain models, like class diagrams, allow the user to model quite informally at first. This is highly desirable in early phases of the modelling process.

Turning to more formal representations, abstract concepts like a class constructor, the formal specification of earlier informally defined attributes and methods and the access operations for the attributes are indispensable. In many cases a software development tool can identify such missing details and ask the user to complete the model. In addition such tools can ensure the consistency of different sub-models (e.g. class and sequence diagrams). Through this process learners understand which components belong to a complete model and how the consistency of different sub-models can be realized. Deeper problems must be identified and resolved in lesson discussions. It would be naive to count on system support to offer solutions to the tremendous variety of potential semantic (logical) problems.

3.3 Visualization of abstract concepts with change of the observation perspective

Modelling tools offer the possibility to change the size of diagrams and the degree of detail. Learners can view the information at or on different levels of abstraction. They can concentrate on the elements that are essential for a concrete problem solution phase. In order to, e.g. analyse the structure of a class hierarchy, the representation of detailed information on single classes (attributes and functions) is not required if inheritance is not the topic of discussion. SDEs offer multiple step execution and undoing of functions. Of course learners expect such application standards in software that is needed for exploration. Since many of the afore-mentioned model check mechanisms only indicate problems but are unable to correct them automatically, learners soon become aware of their own errors. It remains a complicated learning process to understand why an error occurred (Tholander 2001) and even more difficult to derive better strategies for future modelling tasks from understanding the error. The traditional educational process allows for the implementation of correction recommendations worked out by the learning group.

4. DESIGN CRITERIA OF THE DIDACTICS OF INFORMATICS

The concept of exploration is a component of the educational process. Therefore learning theories and the principles and methods of the Didactics of Informatics crucially influence the arrangement of that concept. The learning theory of constructivism (Vygotsky 1978), which is eminently suitable for accommodating various constructive requests of informatics education, is applied. From our examination of the didactics of informatics we have concluded that learners in the context of OOM (Rumbaugh et al 1991) should:

- get to know and design structures (static basic concepts);
- understand and control processes (dynamic basic concepts);
- analyse, modify, construct and assess object-oriented models following the procedure of *select elements, structure, model* in developing examples of use of the everyday world.

Therefore the authors consider the following components and functions of courseware promote exploration.

4.1 Basic concepts on different abstraction levels

In order to give learners with varied pre-knowledge a cognitively demanding but not overtaxing entry to object-oriented modelling, the basic concepts must be capable of exploration at different levels of abstraction. For beginners, representation forms are suitable which should illustrate the basic idea of object-orientation in examples of the everyday world. An animation view on a reality section shows the objects, their attributes and behaviour. Advanced learners are introduced step by step to more formal representation forms such as selected UML diagrams. The meaning and effect of object view, class view, and message exchange view must be explored.

At least one static and one dynamic representation form are necessary to describe the architecture and the course (the time change of the state of an object network). The class diagram as a central structure description of the architecture of object-oriented models is inalienable. To illustrate the dynamics of the message exchange, sequence and collaboration diagrams that emphasise various aspects of the process are invaluable. While methods which exclusively reach their behaviour by communication with other objects can be modelled superbly by means of sequence diagrams, calculation rich methods which have low, or no, object communication can only be partially designed with their aid. For this reason structograms are used within imperative problem solving and a suitable combination of sequence diagrams and structograms must be developed for an exploration. The step from a reality section to a class diagram can be simplified with object diagrams.

It must be possible to switch selected diagrams on or off. The complexity of the diagrams is influenced by the fact that components and details may be hidden or uncovered. In an OOM various processes run, objects become instantiated, links to other objects are built up and removed, messages between objects are sent, objects are destroyed. To be able to understand these processes and to judge their consequences the processes are activated by the learners, watched, recorded and discussed. On a graphic level, independently of a special programming language, objects of a class of a given class structure are created, their methods executed and the changes in the object states monitored.

4.2 Synchronization, transformation and evaluation of diagrams

Synchronization represents a 'didactic bridge' between different diagrams. This makes it possible to bring the single diagrams together to form a picture of the complete system and thus overcome a known cognitive barrier. These may be a set of diagrams that represent the same information under different points of view. As far as it is possible, these views must be synchronized automatically. An example is the representation of a simple reality section, e.g. the move operation on a graphic object, its representation within an animated collaboration diagram and the synchronized representation of the corresponding source code. In addition there is the possibility of recording running processes in the system and rerunning the recording later in different views. It is not always possible or desirable that the exploration system automatically synchronizes views. In order to enable learners to transform representation forms on their own e.g. the transformation of class to object diagrams and vice versa, functions must be provided with which the consistency of views can be checked. For every object diagram a well-defined class diagram can be determined but the reverse is impossible, because an unlimited number of object diagrams can be derived from a class diagram. However, we can check whether an object diagram is a valid instance of a class diagram and we can identify where the structure is damaged.

5. EXPLORATION MODULES AS LESSON MEDIUM

To embed the concept of exploration in the traditional educational process we need to develop suitable learning aids (EMs). With this software collection learners can find out connections and discuss their discoveries during the lesson with the learning group. Interactions between the learners and the EMs are very important. We haven't used learning texts as part of the EMs but good textbooks on OOM are available during the lesson.

EMs have a structure like the technical experimenting box of building blocks. Learners combine solution elements to object-oriented models. They can examine predefined models or design their own models. The diagnosis options of models from the database are extensive. For freely chosen designs a standard set of representations can be used for illustration and exploration. However, an extensive model critique is carried out in the lesson conversation. EMs were developed for learning object-oriented basic concepts and they complement the use of software development tools as a learning aid. The pressure on learners from the syntax and semantics of a programming language in the first lessons of the course is avoided by the use of EMs (Schubert 2000).

To examine a learning object with an EM suitable observation and manipulation components are required. Complex connections, principles and methods are illustrated and exemplified. The observation component manages all diagrams (views), animations and the changing of views. It also is used to generate sub-models automatically (e.g. class diagram to a given object diagram). EMs stimulate learners to develop hypothesis formulation about the object of examination giving them the opportunity to explore the object, manipulate it and check whether changes cause expected consequences. This is possible with all construction and parameterisation processes of object-oriented models. In addition, at the half-automated view transfer (e.g. from class diagram to object diagram or from class diagram to real object) learners must interact with the system and make decisions.

In this context the dialogue component of the EMs is extremely important. It presents learners with the consequences of their actions. In the learning process the two main areas of observation and manipulation are combined. An object of examination is altered and carefully observed. The consequences of the actions are analysed and judged. Following a modified plan observation and manipulation are continued. This process leads to a higher level of understanding.

6. LEARNERS' APPLICATION OF EXPLORATION

Exploration promotes learning-by-doing oriented, problem and situation based learning when correctly integrated in the informatics education (van Weert 2001) The recommendation of learning scenarios illustrates this (Table 1, number of points as indicator for the degree, how much the authors recommend the application in the informatics education).

In object oriented modelling the linking of basic concepts with model elements, selected views of a model and the view change (respectively the combination of different model views) requires an explorative approach. In all cases the abstract connections are processed in a way that makes interaction with the concepts possible. The observations of the learners are embedded in the theory of object-oriented modelling in the informatics lesson and therewith turn to knowledge.

Table 1. Application of exploration in Informatics education

Lesson phases	Model elements	Model view	View change	Informatics system
Problem orientation	••••••	••••	••••	••
Motivation		••••	••••••	
Introduction to new material	••	••	••	••
Exercise/application	••	••	••	••
Subject specific transfer		••	••••••	
Controls of learning success			••	

However, exploration is not equally effective in all areas of OOM and all the afore-mentioned lesson phases. We must have realistic expectations of exploration. Our research shows that exploration can be particularly recommended for problem orientation in the fields of model elements, model view and view change. Using specific visualisation and interaction possibilities this is especially valid for single model elements. In the lesson phases concerned with motivation and subject specific transfer, the view change of the learners is promoted due to computer supported synchronisation of views and their visualisation. When learners discover the meaning of a view change in an example world they are motivated to work with this method (Table 1).

The EMs have a double pedagogical function, because they can either be used as learning aids or can be analysed and discussed as an example of a complex software system. The discovery progress of the learners is tied to general conditions e.g. the establishment of quality demands as well as inclusion in teacher education. Student teachers of informatics on the one hand have to be prepared to use exploration as a new teaching method and on the other hand to see EMs as learning aids.

In the process of the development of the 'Didactic System for OOM' the following steps are the next to be explored. The activities of learners and concrete examples for lesson phases with EMs will be worked out. A group of 11 students of Informatics at the University of Dortmund will implement an exploration environment for selected object-oriented basic concepts until July 2002 and the component 'Exercise Classes' (Brinda and Schubert 2001) of the didactic system will be applied and empirically evaluated in student teacher education of informatics in spring 2002.

REFERENCES

Appelrath, H.-J., Boles, D., Claus, V. and Wegener, I. (1998) *Starthilfe Informatik.* B.G. Teubner, Stuttgart.

Brinda, T. and Schubert, S. (2001) Didactic system for object-oriented modelling. In *Networking the Learner*, Proceedings of the seventh World Conference on Computers in Education (WCCE 2001), Kluwer Academic Publishers, Boston, pp 473-482.

Rumbaugh, J., Blaha, M., Premerlani, W., Eddy, F. and Lorensen, W. (1991) *Object-Oriented Modelling and Design.* Prentice-Hall, New York.

Schubert, S. (2000) The impact of modelling in informatics education on collaborative learning with school Intranets. In *The Bookmark to the School of the Future*, P. Hogenbirk and H. G. Taylor, (eds.), Kluwer Academic Publishers, Boston, pp. 247-258.

Steinkamp, D. (1999) *Informatik-Experimente im Schullabor.* Diploma thesis, Dept. of Informatics, University of Dortmund.

[http://ddi.cs.unidortmund.de/lehre/diplom/steinkamp/]

Tholander, J. (2000) *Students Interacting through a Cognitive Apprenticeship Learning Environment.* Euro-CSCL 2001, Maastricht.

[http://www.mmi.unimaas.nl/euro-cscl/Papers/162.pdf]

Turner, A. J. (1998) Trends in teaching informatics. In *Informatics in Higher Education*, F. Mulder and T. van Weert, (eds.), Chapman and Hall, London.

van Weert, T. (2001) Co-operative ICT-supported learning. A practical approach to design. In: *Informatikunterricht und Medienbildung*, R. Keil-Slawik and J. Magenheim (eds.), Köllen, Bonn, pp. 47-62.

Vygotsky, L. S. (1978) *Mind in Society.* Harvard University Press, London.

Learning Software Engineering with EASE

Dirk Draheim
Institute of Computer Science, Free University Berlin, Takustrasse 9, 14195 Berlin, Germany

draheim@inf.fu-berlin.de

Key words: Didactics of Informatics, Higher Education, Teamwork

Abstract: The paper describes EASE - an ultra-lightweight software engineering process model that specifically targets student projects in higher education. The model has been obtained by analysing the requirements of student projects and then carefully combining relevant concepts from state-of-the-art software management concepts. Recommendations are given for structuring the course as a whole, as well as managing the work. The latter is oriented towards mature andragogical methodologies. EASE has significantly improved the learning outcomes in both undergraduate and graduate courses and the paper proposes that it should be adapted and trialled in secondary education and industrial training environments.

1. INTRODUCTION

EASE (Education for Actual Software Engineering) is a methodology for the teaching of software engineering skills in higher education. It applies to projects, i.e. every kind of course in which students build software systems by doing teamwork. Such projects belong to every core curriculum in computer science, both undergraduate and graduate. EASE consists of the following components (Figure1):

- driving forces;
- a process architecture;
- a practices catalogue.

In a broad sense the *driving forces* define the problem the lecturer is faced with. The *process architecture* and the *practices catalogue* together constitute the solution. The practices form a *process pattern language* (Coplien 1995) (Figure 2).

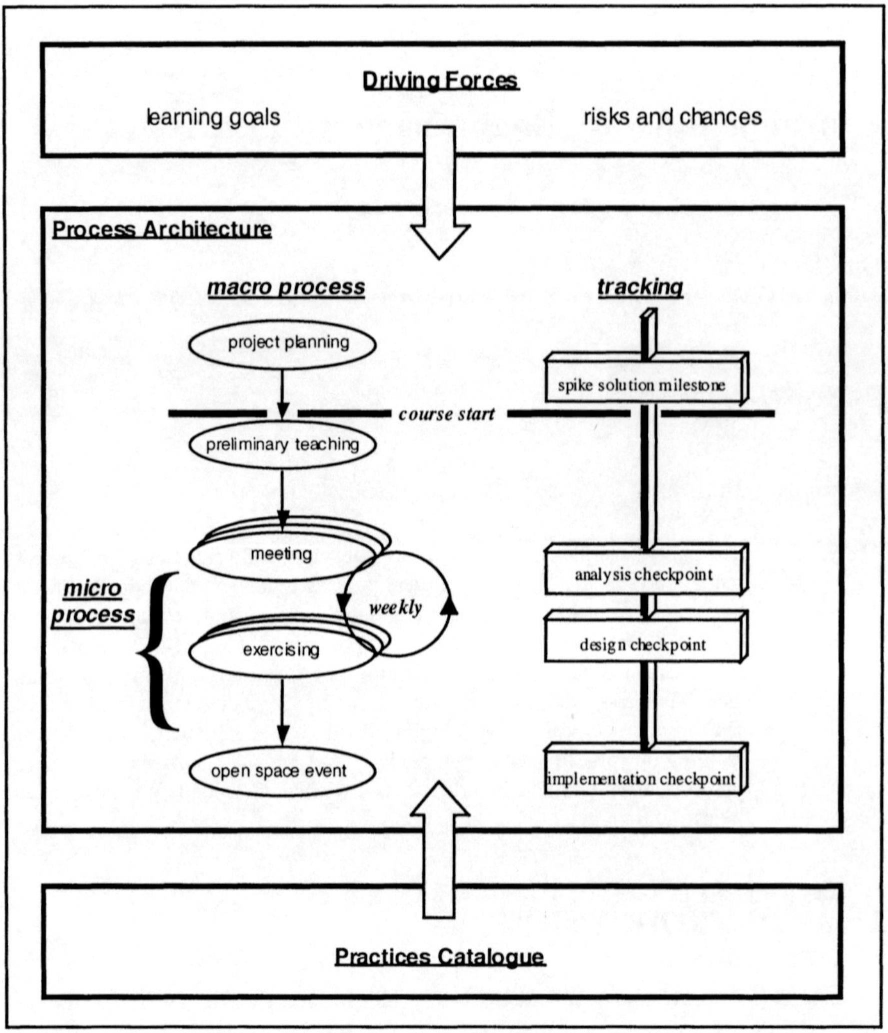

Figure 1. EASE - Education for Actual Software Engineering

EASE is the first process model that has been designed from scratch to accommodate the specific needs of education. Wherever appropriate EASE is oriented towards concepts found in today's software engineering process. With respect to didactics, EASE is influenced by collaborative learning (Brufee 1983).

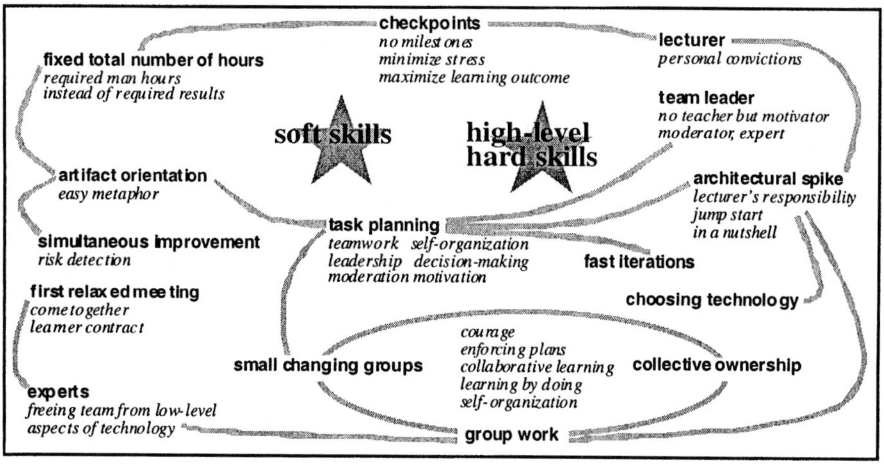

Figure 2. Practices pattern languages

2. DRIVING FORCES

We consider learning goals and risks and chances (or opportunities) are the driving forces of a software practical course. It is noteworthy that these driving forces are fundamentally different from those found in practice. Software management is a product-oriented task. The main driving force is risk and the entities to deal with during risk estimation are different from those in education - namely quality, quantity, cost, time and productivity.

A software course inherits risks and chances from adult education. There are also specific risks with regard to software engineering. Different individuals perceive the same topic in totally different ways (Varela 1988; 1993). Participants have different dispositions and hold different opinions about the purpose of computer science. As a result, the students have different objectives concerning their participation in the course. Like Siebert (2000) we regard this point as crucial for adult education and believe that the course can hardly be successful if it is ignored. However, we believe that is not only a risk factor but also a chance as different skills and opinions may contribute to collaborative learning (Brufee 1983).

These are chances. Students are receptive, hard-working and well-motivated. There is the risk that motivation may vanish. We believe it is very important to respect the students' different points of view in order to keep motivation high.

In nearly every software project, we observed students who only wanted to gather know-how about technology and were not interested in other learning goals. They have an attitude similar to that of real programmers and sometimes "baby duck syndrome" occurs. Even more problematically, these students often tend to be non-communicative. They are not willing to work together with students that do not exactly share their opinions. In the worst case a project team consists of several small competing groups of programmers (and a bunch of increasingly intimidated other participants). With EASE's micro process, these students quickly become major contributors to teamwork.

Lecturers may introduce other risks when they simplify complex software engineering concepts. Sometimes concepts are simplified to such an extent that students are misled. Take object orientation as a prominent example - often is taught as a modelling the world with objects paradigm, without caveats concerning the new impedance mismatch between problem domains and solution domains of non-trivial systems, that is the result of object orientation.

EASE is designed to mitigate risks and exploit chances. Therefore EASE fosters teamwork and collaborative learning, self-organization and learning by doing.

Currently EASE uses its own, quite coarse, taxonomy of learning aims. These include *hard skills* which are:

- technology-related, low-level: detailed know-how about concrete technologies, e.g. programming languages, APIs, operating systems, CASE tools;
- technology-related, high-level: the ability to recognize advantages and disadvantages of concrete technologies with respect to a current ambiguous task;
- method-related, low-level: know-how about concrete software engineering techniques and concepts, e.g. specification languages, modelling languages, process models, design patterns, GRASP patterns;
- method-related, high-level: the ability to appreciate the respective importance of concrete software engineering techniques and concepts.

There are also *soft skills*:

- courage, skills concerning teamwork, leadership, decision-making, enforcing plans, motivation, and moderation.

Finally there are *key skills*:

- receptiveness, efficiency, ability to learn.

EASE is supposed to support the learning of soft skills and high-level hard skills, but it does not prescribe concrete learning goals. For instance, in a particular project modelling, specification, design, or even coding may be emphasized. EASE provides a framework for learning and exercising (or applying) software engineering. The lecturer defines the learning goals.

3. PROCESS ARCHITECTURE

The core process of EASE has a sophisticated, yet easy to learn architecture. It consists of a macro process that is accompanied by tracking (Figure 1). The macro process starts with a project planning phase and is followed by a preliminary teaching phase. Then a controlled iterative micro process is entered, which consists of repeated steps, namely meetings and exercises. The micro process takes the majority of the project's time. The project is closed with an open space event. Tracking encompasses a spike solution milestone and checkpoints for analysis, design, and implementation.

EASE's meeting concept was developed by adapting the iteration planning of Extreme Programming (Beck 1999), and it is inspired by Collaborative Learning (Bruffee 1983) and Action Learning (Revans 1982). The resultant micro-process iteration bore a striking similarity to the session workflow of Entraînement Mental (Chosson 1975). We regard this as a further justification of our approach. EASE, in common with the USDP (United Software Development Process) (Jacobsen et al 1999) has a controlled simultaneous improvement of artifact sets, termed two-dimensional process structure in USDP.

3.1 Macro Process

Project Planning

Some concepts in EASE presuppose that the course has enough participants to be divided into several teams, so that each team has ten to forty members. Each team has a lecturer or a tutor as their leader. With minor changes EASE can be used in courses of smaller scope.

The lecturer prepares the course defining learning goals based on personal convictions. Learning goals may be predefined because of an encompassing study plan. The lecturer defines the desired system as a vision document. Each team is set the task of realizing as many features as possible. Work is finished when every student has worked a fixed total number of man hours - a figure agreed at the beginning of the course. The course is finished with a checkpoint - not by a milestone as that enforces a predefined feature set. The lecturer defines sets of simple artifacts - EASE task planning is artifact-oriented. These are divided into an analysis set, a design set and an implementation set - an intentionally coarse division. In contrast to the theory of stage-wise or waterfall-like software management, EASE assumes that artifact sets are developed simultaneously from the outset - simultaneous improvement. EASE is deliberately not prescriptive at the level of engineering activities in order to avoid micro management. Instead it is designed to be an instructive challenge for the students. They must find the tasks that contribute best to progress on their own.

The definition of the artifact sets is accompanied by choosing technologies and techniques and by developing a system architecture. For this purpose the lecturer has to develop an architectural spike. This is a prototype system which realises just one or a few representative use cases of the desired system. All the technologies on which a chosen system architecture is based are used. There are subtle differences between the reasons for architectural spikes in EASE and those found in practice. The purpose in practice is for estimating risks associated with a particular architecture. In EASE the resulting spike solution is used to present the chosen architecture and technologies to the students in preliminary teaching. It serves as both start point and reference point during the rest of the course. The spike solution must consequently avoid encompassing unnecessary features like exhaustive functionality, constraint checking, and fancy user-interface layout. Using a spike this way has the following benefits:

- "in a nutshell" - the students understand the architecture as a whole.
- "jump start" - the students have early success in handling technology.

In this version of EASE, the students do not develop the architecture. This is considered too challenging. In practice it is a senior architect's task. The lecturer fixes days for the analysis and the design checkpoint.

Preliminary Teaching

In a first plenum the desired system is explained. EASE is explained, especially the micro process. An overview of the architectural spike is given. The students are assigned to teams randomly. Homogenous groups should be avoided. An exception may be made if there are requests for homogenous female teams. The students of each team get together in a first relaxed meeting and a first simple task is assigned to the students.

Micro Process

Each team enters the micro process, which is fast iterative and consists of alternating weekly meetings and exercises. Meetings are for task planning. During exercise periods the tasks defined in the meeting are carried out. At each meeting the whole team and its team leader get together. This team leader is the motivator, as required a moderator, occasionally an expert.

A meeting takes approximately one and a half hours. During the meeting five well defined activities take place (Figure 3). It is important to note that in the first few meetings these activities are viewed as sequential steps that define a workflow. Once students are used to the meetings activities should be carried out in parallel. In this way, the mutual dependencies between the activities are exploited to improve the outcome. However, the workflow technique may be used if teamwork gets stuck. For better understanding activities are presented as workflow steps in this paper and that is the way they should be explained in preliminary teaching.

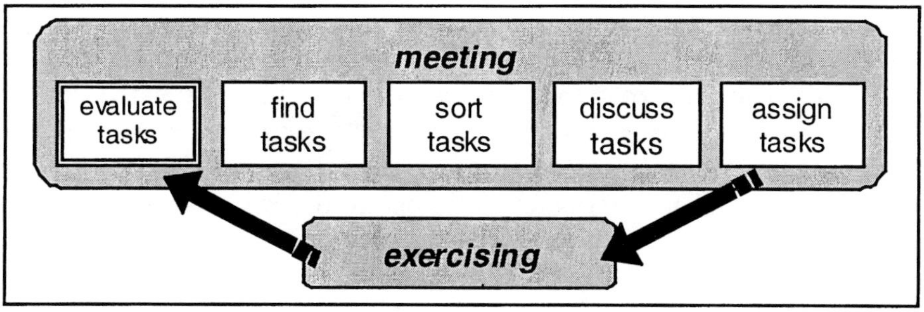

Figure 3. EASE - micro process

First, the results of the previous week's effort are discussed and validated. Then the team concentrates on finding new tasks. Task planning is artifact oriented. The main question is - what has to be done next in order to contribute most effectively to the growth and the quality of the artifact set? Brainstorming techniques are very helpful. All suggestions must be taken seriously and gathered for further discussion. Usually more work is found than is possible to do during the following week so tasks have to be sorted with respect to importance. Based on their current knowledge and opinions the tutees discuss what to do next and why. This results in a task list, ordered by importance. Students learn to appraise the relative importance of the mutually dependent families of software engineering activities.

Alternative solutions for the tasks are suggested, discussed and finally one of them is chosen for each task. Tasks that are too large to be carried out by a group in a week have to be split into several tasks. In the end every task is assigned a new, succinct title and a short, nonetheless precise, description. Tasks then have to be assigned to team members. Usually a task is assigned to a group of two to four participants. Group members estimate the time needed to carry out the task and agree the time contribution of each member. Tasks that cannot be assigned are considered the following week.

The team is divided into small groups that work on tasks. This division is not static and at each meeting, new groups are formed. The tutor has to monitor the selection. In a project threads of strongly dependent tasks exist. If a task has a predecessor, it should be assigned to the predecessor's group but at least one group member should be replaced with a new team member. This concept of small and frequently changing groups is crucial, because we pursue a generalized notion of collective ownership. It is the target of the exercise that students learn about the overall structure of the system that is built up. Ideally, every student should have an understanding of every technology and concept used in the project, and how they interact. Consequently students do not over-specialize and the learning outcome is shifted in the direction of high-level hard skills and soft skills.

The meeting is followed by exercises. Tasks are carried out using the teamwork approach. Often the group work can be further sub-divided into smaller tasks for every group member - nearly always true for coding tasks. Even with this sub-division the group should stay together in the same room, although every team member may be working on their own. If a problem arises all group members immediately work together to find a solution. Group members help each other on demand. If the task is coding, pair programming is recommended. If a new group member joins every question has to be answered patiently. The concept of small changing groups makes tutees learn from each other. At the next meeting results are evaluated. Students talk about their experiences, describe problems that occurred and explain how long they worked (they have to work a fixed total number of hours). Independent of the total outcome of work we find students typically are willing to respect the work of other, perhaps less experienced, students. Our approach tackles the problem that students often do not trust the industry of their colleagues because they have different capabilities - a potential source of squabble. The overall teamwork should be supported by appropriate simple tools like repositories, mailing lists and discussion forums. The course terminates with *an open space event.* This realises the implementation checkpoint. Each team explains its system to the other teams in a detailed plenum presentation. Then all students gather in computer rooms to examine and test all the systems.

3.2 Tracking

On the fixed days of the analysis and design checkpoint, the respective artifact set is reviewed by the team leader. In a plenum, results are presented briefly, compared and discussed. We use the term checkpoint instead of milestone. A milestone defines a targeted result. A checkpoint is a softer control mechanism and fixes a day for reflection and problem exploration.

3.3 Case Study

We developed EASE based on the experiences gathered in our lectures on software engineering and their satellite projects. EASE was thoroughly used for the first time in summer 2001 in the lecture "Softwarepraktikum" (undergraduate, 4 semester hours) at Freie Universität Berlin. It significantly improved the learning outcome. It was used by three teams each consisting of fifteen to twenty tutees. Only eighty per cent of the participants studied computer science. Ten per cent had extensive skills in net programming or using databases, usually only low-level hard skills. At the beginning of the course the other students had hardly any knowledge about distributed systems or persistent data. By the end of the course every student could deal with the technologies and several complex notions shown in (Figure 4).

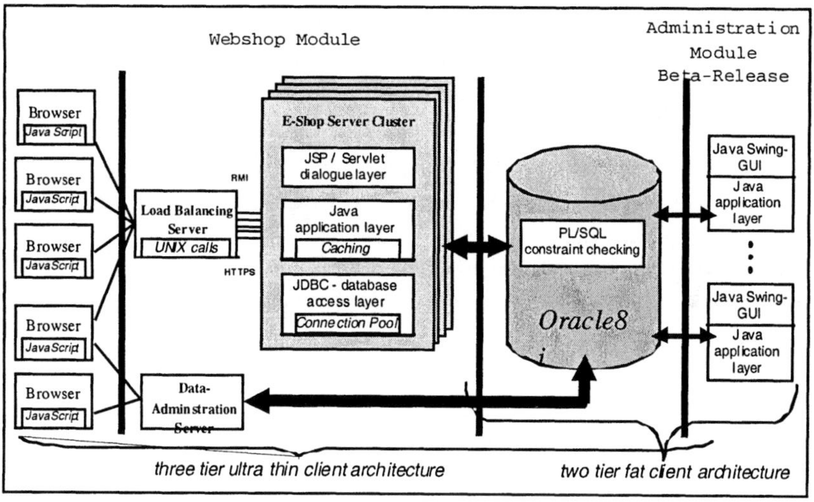

Figure 4. Case study - system architecture

The artifact sets encompassed a web user interface prototype as the result of storyboarding, a UML data model, a textual requirement specification, deployment and design diagrams, commented code and automated tests. The students learned using SourceForge's open source management capabilities, Concurrent Versioning System software, modelling tools, Integrated Development Environments and GUI builders.

4. FURTHER DIRECTIONS AND RELATED WORK

We recommend EASE should be adapted to other areas, e.g. secondary teaching or industrial training. In a graduate course it should be possible for students to take over the project-planning phase and are evaluating this in a challenging practical course on EJB (Enterprise Java Beans) application servers. A web-enabled, team-oriented EASE project management tool is under development. Lessons learned from using XP and the Rational Unified Process (RUP) in student projects are presented in a number of papers (Lippert et al 2001; Traub 2001). In our opinion it is not possible to use XP in a student project. Teaching XP using just a few XP notions misses the point. It is a common misunderstanding that XP is an ultra-lightweight process model. Instead it defines a highly sophisticated net of practices that can be exploited only in a real world project. Unsophisticated attempts to use XP in education can seriously mislead students (nonetheless XP can be taught - by using usual "frontal" teaching). This applies even more so for the heavyweight RUP (Kruchten 1999). The Unified Process for education UP/EDU (Robillard 2001) is the result of reducing RUP so it can be used in education. The result is still overpowering as UP/EDU compels the lecturer to teach certain topics, particularly subtle proprietary RUP notions.

If a professional process model is used in education naïvely or in an oversimplified way, its constituting concepts are probably not exploited at all. A student project differs from a real world project in scope, in driving forces and in organisational culture. EASE is a reusable software engineering process for practical courses in higher education. It is ultra-lightweight and combines new concepts with proven concepts from software management and didactics. EASE targets the learning of soft-skills and high-level hard skills.

REFERENCES

Beck, K. (1999) *Extreme Programming Explained - Embrace Change*. Addison Wesley.

Bruffee, A. (1983) *Collaborative Learning: Higher Education, Interdependence, and the Authority of Knowledge*. Johns Hopkins University Press.

Chosson, J.-F. (1975) *L'entraînement mental*. Le Seuil.

Coplien, J. O. (1995) A Development-Process Generative Pattern Language. In *Proceedings of1st Pattern Languages of Program Design*, J. O Coplien and D. C. Schmidt (eds.), Addison-Wesley, pp. 183-237.

Jacobson, I., Booch, G. and Rumbaugh, J. (1999) *The Unified Software Development Process*. Addison-Wesley.

Kruchten, P. (1999) *The Rational Unified Process*. Addison-Wesley.

Lippert, M. et al (2001) XP lehren und lernen. In *Software Engineering im Unterricht der Hochschulen*, H. Lichter and M. Glinz (eds.), dpunkt.verlag.

Revans, R. W. (1982) What is Action Learning ? The *Journal of Management Development*, Vol. 1, No. 3, pp. 64-75, MCB Publications.

Robillard, P. N., Kruchten, P. and DíAstous, P. (2001) YOOPEEDOO (UPEDU): A Process for Teaching Software Process. In *Proceedings of the 14th Conference on Software Engineering Education and Training*. IEEE.

Siebert,H. (2000) *Didaktisches Handeln in der Erwachsenenbildung - Didaktik aus konstruktivistischer Sicht*. Luchterhand.

Traub, S. (2001) Einsatz von objektorientierten Technologien und Softwareentwicklungs-prozessen in der Lehre. In *Proceedings of NetObjectDays 2001*. TranSIT.

Varela, F (1988) *Cognitive Science - a Cartography of Current Ideas*. Pergamon Press/Leuven University Press.

Varela, F. (1993) *Kognitionswissenschaft - Kognitionstechnik*. Suhrkamp.

Object models of IT-systems supporting cognitive structures in novice courses of Informatics

Peter Hubwieser
Department of Informatics, Technische Universität München, 80290 München, Germany

Peter.Hubwieser@in.tum.de

Key words: Basic Concepts, Cognition, Curriculum Policies, Informatics, Secondary Education

Abstract: Designing a new mandatory subject of informatics for the 6th grade of the Bavarian Gymnasiums we decided to use a certain type of object modelling to bridge the gap between two almost opposite objectives. The first was to enable the students to use IT-systems in a competent and autonomous way - the second to teach durable and transferable basic concepts. While some criticisms have been levelled against the benefits of mental models, we rely on these mental models to support our novice students in applying and mastering IT-systems. Our views are supported by some recent results from pedagogical psychology. In this paper we present a suitable modelling technique that will be used in the novice course of the new subject and some results of the first evaluations of the preliminary pilot courses.

1. INTRODUCTION

Following years of confused discussion whether informatics in schools should be implemented in the form of a mandatory, dedicated subject or integrated within other subjects like Mathematics, Languages or Science, the conclusion that it is necessary to combine these policies is gaining acceptance in Germany. This is caused by three main factors:

- It is absolutely necessary that nearly all school subjects should make use of the opportunities that are offered by e-learning.

- In nearly every subject the consequences of the information society on the specific subject area are discussed.
- In the near future virtually all students will have free access to a computer (currently 97% of 11 year-olds have access) with a rich software resource.

Students have demonstrated that they are able to acquire basic user skills without guidance (Mitra and Rana 2001). It is therefore no longer necessary to offer them just access to hardware and software systems or to drill basic user skills, but to teach them how to use IT-systems competently and help them answer the questions that are posed to them during the use of these systems:

- How does that machine work?
- What are the basic concepts involved in its construction?
- What are the limitations of its use?
- How can I really control its affects?

The only possible reaction to these insights is to make considerable use of IT-systems, to reflect and discuss their consequences throughout the curriculum, to teach basic concepts that are essential for a real understanding of the working principles of systems and master their usage within a dedicated subject called "informatics".

2. A NEW SUBJECT CALLED "INFORMATICS"

As a result of these considerations Bavaria decided to introduce, from 2003, a new, mandatory subject "informatics" in the 6th grade at all its 400 Gymnasiums. Each year this will involve about 50,000 eleven to twelve year-old students. Currently we are testing the concept with regard to content and teaching methods at 20 schools, involving about 40 classes and 1200 students.

This new subject has to meet the demands of a broad variety of different social groups like companies, universities, parents, students and teachers. Long and intensive discussion beneath these groups showed there should be two central objectives which, at first glance, appeared to be diametrically opposed:

- the students should be taught the *application* of certain specific and wide-spread IT-systems (graphic and word-processors, web editors, e-mail clients) to real-life situations;
- the acquired skills and concepts should be *valuable for a very long time*, even after their period of education, and should be transferable over a broad variety of systems.

We tried to satisfy these demands by offering a teaching concept that stimulates the students to explore long-lasting structures (which might serve as mental models to support real understanding) as well as encouraging them to work extensively on documents using a set of modern, well-known standard software systems.

3. MENTAL REPRESENTATIONS

With regard to mental models we have to ask ourselves if human beings really use a situation-independent mental representation before the application of any tool. This seems obvious from a common-sense rationalistic point of view but there are strong arguments against this, notably the criticism of Winogard and Flores (1986) who quote the philosophy of Heidegger and the experimental results of Manturana.

Heidegger claims that a highly skilled craftsman like a carpenter usually is not aware about the hammer he is using. Instead he has a "readiness by hand" attitude towards the hammer, which operates outside his consciousness until the hammer is not where it is expected to be or not working properly. This causes "breakdown" and finally draws the attention of the craftsman to the tool.

Maturana investigated the retina of frogs and detected that their perception, e.g. of a fly, was not based on a formal representation of the fly but emerged as a local variation of light compared to the surroundings. The result inspired Maturana to his radical constructivism, regarding all living things as closed "autopoietic systems". The adaptation of these radical constructivistic biological concepts to the context of human learning contributed to the development of moderated approaches, such as the "situated cognition concept" (Gerstenmaier and Mandle 1996). This assumes human mental models, which result from the learning process, are not only dependent on the taught concept, but also on the situation and the context in which they were acquired.

Nevertheless we are convinced that mental representations are used during the application of a tool under at least one of the three following circumstances:

- the user is a novice concerning the tool;
- the tool has a very complex structure;
- the tool is used in a very complex application domain.

All these circumstances are satisfied when eleven year-old students are learning to use and apply standard software systems. As Winograd and Flores (1986) conclude "Human cognition includes the use of representations, but it is not based on representations."

We claim that under the circumstances described above mental models (regarded as representations) are at least helpful. Consequently we wish to identify which kind of mental models will help the students to use the systems most efficiently. Results from modern pedagogical psychology and recent developments of software engineering models offer help in this identification.

4. MODELLING TECHNIQUES

4.1 Cognitive Mapping Techniques

During the past decades a variety of mapping techniques has been explored in the field of pedagogical psychology. The techniques were used to enhance learning processes, to support the revision of concepts or strategies and to foster the acquisition of cooperation skills or analyse the modelling of learning processes (the last by enabling qualitative measures of learning results).

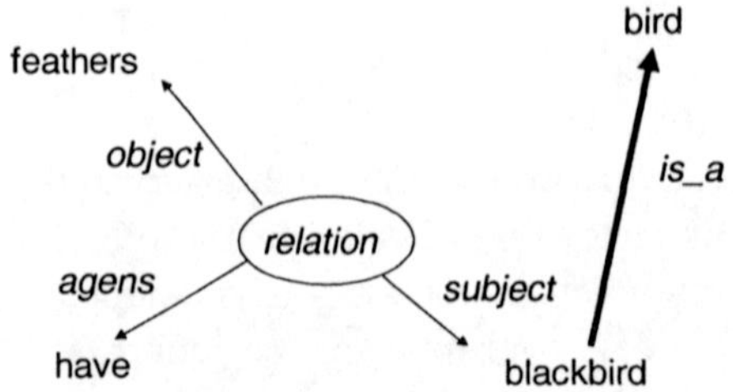

Figure 1. Propositional network

One of the first generally accepted techniques was the concept of Propositional Networks (Figure 1) (Rumelart et al 1972). This was originally designed to graphically represent the structure of human cognition. The graphs consisted of two kinds of nodes (concepts and relations) and edges which symbolized subject, object and agents in the context of relation-respective generalization ("is-a" relation). Based on this a number of techniques followed, such as Concept Maps (Jüngst 1992) and Expert Maps (Boothby and Alvermann 1984). The former use nodes to signify concepts and additional information (e.g. restrictions) and marked edges to assign relations (e.g. generalisation and aggregation - Figure 2). Expert maps use nodes to symbolize concepts and constraints and unmarked, but weighted, edges to represent different types of connections between the nodes (Figure 3).

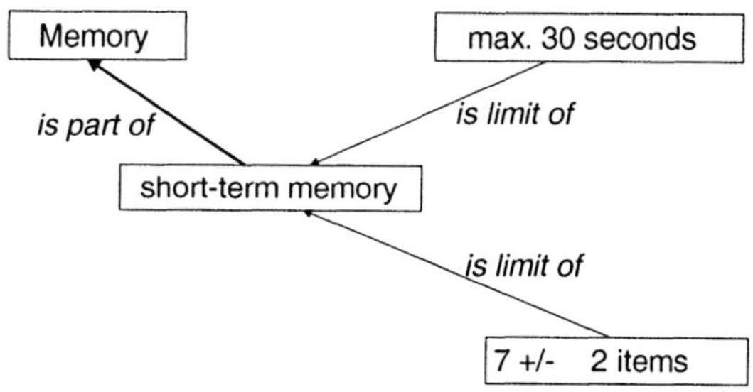

Figure 2. Concept map

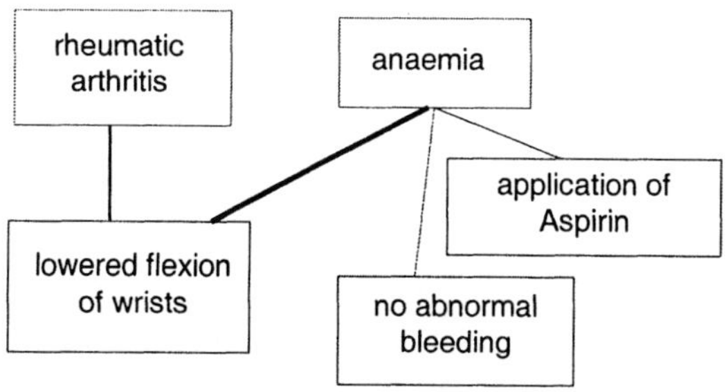

Figure 3. Expert map

The application of these graphs was intensively evaluated and showed some remarkable results:

- concept mapping techniques improved the memory capacity compared with textual representations;
- expert map sequences supported the application of concepts;
- concept mapping techniques supported the coordination of conceptual and numerical knowledge during the application of physical formulae (Mandle and Fischer 2000).

4.2 Software Models

Simultaneously with the development of mapping techniques by pedagogical psychologists graphical modelling techniques were developed in software engineering, culminating in the cooperative design of the current de facto standard, the Unified Modelling Language, by key protagonists of Object Oriented Software Modelling (Booch et al 1997). This was not a fortuitous correspondence - these object models were a very specific adaptation of psychological Concept Maps to the needs of software developers and their customers. The success and broad acceptance of these techniques demonstrate that they are very sustainable for supporting communication about IT-systems. Parts of them may be applicable within our teaching concept.

5. CONCEPTION OF THE 6TH GRADE

5.1 The Model Construction Kit

Impressed by the success of these mapping and software modelling techniques we decided to base our concept for the new Bavarian subject of informatics on graphical object-oriented models of ICT-systems. These are very similar to the mapping models used in psychology and to the models of the software developers.

Taking the age of the students into consideration, we designed our Model Construction Kit in a similar way to a box of building blocks. It contains the following concepts that, at the end of the course, enable the students to model a Hypertext structure or a roboter:

- objects, classes, attributes, methods and messages;
- relations between objects and classes; aggregation and other associations;
- structures of algorithms; sequence and repetition of instructions, conditional instructions.

Specific notations are used to describe these concepts:

- object diagrams symbolizing structures of real objects;
- class diagrams describing structures of "abstract" classes as descriptions of real objects;
- formal pseudo-code to ease compaction and exact communication.

5.2 Introduction step by step

Starting with the elaboration of the basic concepts ("object", "attribute" and "class") in the context of vector graphics (Figure 4), we develop the models by introducing the concept of aggregation (e.g. paragraphs containing characters) in the context of text processing.

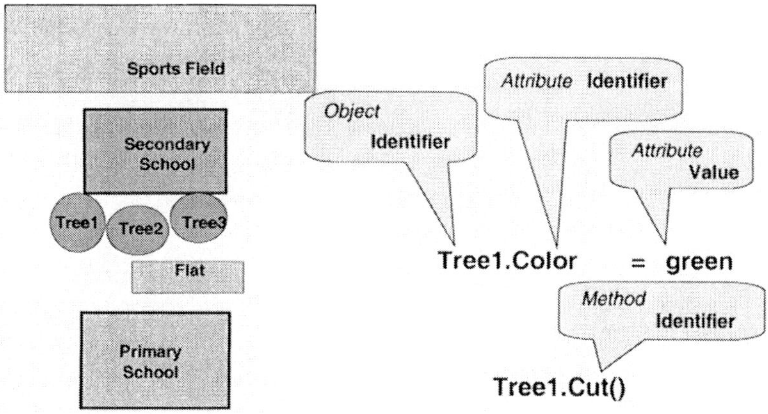

Figure 4. Objects of vector graphics

Subsequently the students encounter the first recursive aggregation, presented by the example of folders that contain other folders (Figure 5).

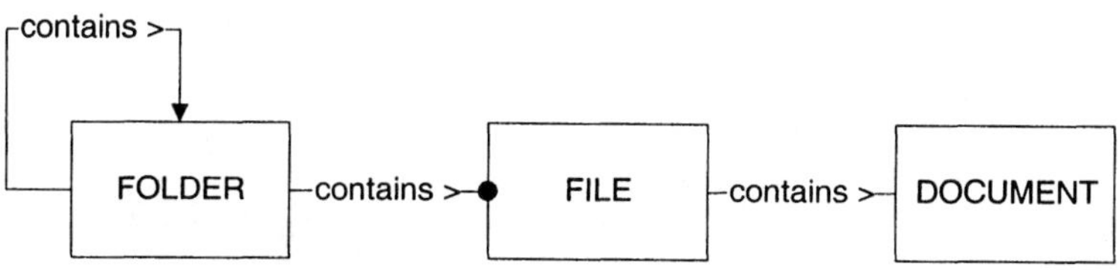

Figure 5. Class diagram - folders containing folders

This enables the development of folder trees, used to represent hierarchical structured information. Later the graphs are extended by the concept of links in the context of hypertext systems. This permits the construction of cyclic graphs (Figure 6). Finally we start to program our own methods using a virtual roboter using "built-in methods" like "forward", "turn left" and so on (Figure 7).

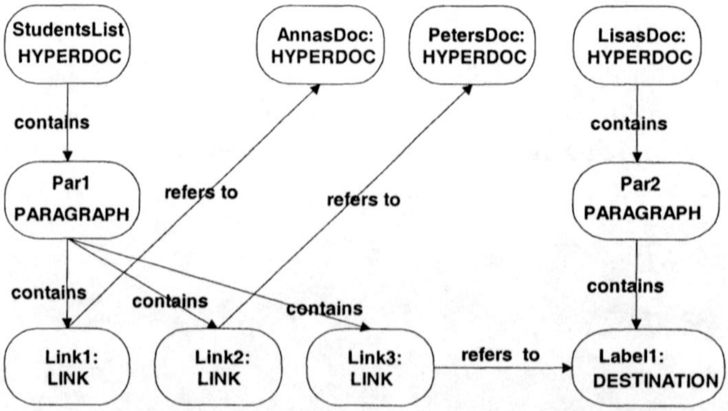

Figure 6. Object diagram - hypertext structure

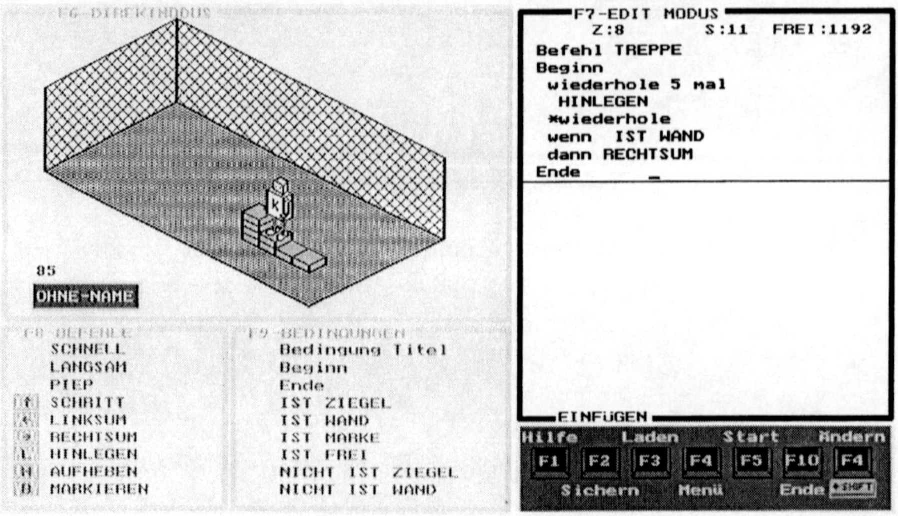

Figure 7. Roboter

5.3 Teaching methods

We emphasize the use of constructivist and integral learning approaches as much as possible. This includes extensive usage of hardware and software systems to construct and edit a variety of documents, representing information and analysing its structures. Visual and haptic illustrations are also considered very useful - for instance the use of wooden stick sets to construct folder trees or the symbolization of an hypertext graph by connecting papers (documents) by woollen threads (links) on a pin-board (Figure 8). All new concepts are introduced by reflecting and analysing the products of the students - drawings, hypertexts, e-mails or self-made roboter programs.

Modern constructivist teaching approaches are supported - for instance:

- *Cognitive Flexibility* by using multiple representations of the same structure (pseudo-code, object model, class model and the document itself on the software system).
- *Narrative Anchors* by integrating the student tasks with thrilling tales.
- *Cognitive Apprenticeship* by working in authentic office environments and using professional software development techniques.

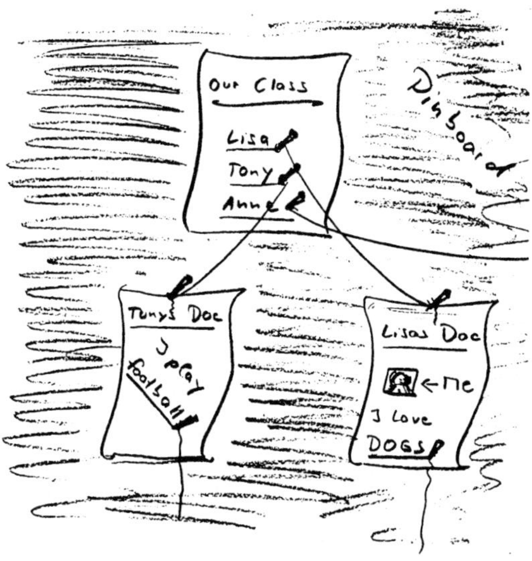

Figure 8. Visualisation of hypertext structure at a pinboard

5.4 The Software Toolkit

First of all we apply a very simple graphic processor. This should be supplemented by use of a pixel-oriented painting program. For text structuring a common word processor will suffice. The choice of a suitable hypertext editor is more difficult. Students should not be forced to learn the syntax of Hypertext Markup Language - they should be introduced to a "what-you-see-is-what-you-get" tool with not too many features. If possible, the software should offer a graphical representation of the hypertext structure. Finally the algorithmic structures require a virtual roboter system such as "Karol the Robot" (http://www.schule.bayern.de/karol).

5.5 More than user skills

Using our methodology we enhance the information processing skills of the students and enable them to understand the basic structure of the systems they use.

We certainly stimulate them to think about the basic structures of the documents they produce (which are very long lasting) in contrast to remembering sequences of product and time dependent terms associated with a specific user interface e.g. menu-entries, checkboxes or hot-keys. The benefits of this principle, compared with the usual drills of pure user skills on a specific system, can be illustrated by the following:

- **Multiple calls**. "Copy", which is common to objects of many different classes, might be invoked by activating a sequence of menu entries (Edit - Copy), by clicking on a symbolic icon on a task bar (a pair of scissors) or by simultaneously pressing a combination of keys (ctrl - c). The result is always the same.
- **Documents and files**. Careful differentiation between the classes "document" (whose objects are located in the working memory of an application) and "file" (whose objects are located on persistent data storage devices) enables students to understand the process of saving and opening a document. "Save" is a method associated with the class document, whose calls are automatically redirected to calls of "save as" as long as there is no file name for the document. However, "open" is (originally) a method of the class "file" (possibly calling a corresponding method of the contained document) which uses the name of an application as a parameter and allows us to open a file (using the file manager) before opening the corresponding application.
- **Clipboard**. The interpretation of a clipboard as an object that may contain every other object that is part of a document makes it easy to describe the working principles of the methods "cut", "copy" and "paste". The call of an objects method "copy" results in copying itself into the clipboard. This enables the call of the copied objects method "paste", which results in an insertion into the active document.

6. THE FIRST RESULTS

After the completion of the first year of the experimental 6th grade, teachers of 28 classes asked their 728 students (62% boys, 38% girls) some questions to investigate their attitude towards the new subject. Simultaneously the teachers had to answer other questions concerning the completion of the syllabus and the software which had been used. Most importantly 81% of the students liked the subject - only 14.6% didn't. Compared to mathematics 58.5% considered informatics was easier, 19.9% comparable and 8.1% felt it was more difficult.

Concerning the syllabus only one of the 28 classes skipped a software topic (hypertext), while four classes did not use object diagrams. All other classes covered the syllabus in full.

Simultaneously the University of Dortmund carried out video-supervision of the project (Hubwieser et al 2001). This confirmed that the object-oriented modelling technique served as a bridge to connect previously known concepts with new ones. It also showed that the students, as a consequence, used the correct technical terminology.

7. CONCLUSION

In our opinion this teaching concept is transferable not only to other lower grades in secondary schools but might, in part, be applied to user training courses on standard software. These often show a deficit in teaching structure and thus lack durability and generality

REFERENCES

Boothby, P. R., Alvermann, D. E. (1984) A classroom training study: Effects of graphic organizer instruction on fourth graders' comprehension. *Reading World*, 23, pp. 325-335.

Booch, G., Rumbaugh, J. and Jacobson I. (1997) *Unified Modelling Language - Semantics and Notation Guide 1.0*. Rational Rose Software Corp., San Jose, California.

Gerstenmaier, J. and Mandl H. (1995) Wissenserwerb unter kognitivistischer Perspektive. *Zeitschrift für Pädagogik 41*, pp. 867-888.

Hubwieser, P., Humbert L. and Schubert S.(2001) Evaluation von Informatikunterrich. (Werkstattbericht). In *Informatikunterricht and Medienbildung*. INFOS 2001, GI-Fachtagung "informatik und Schule", R. Keil-Slawik and J. Magenheim (eds.), Paderborn. GI-Edition Lecture Notes in Informatics. Bonn, pp. 213-215.

Jüngst K.L. (1992) *Lehren und Lernen mit Begriffsnetzdarstellungen. Zur Nutzung von concept-maps bei der Vermittlung fachspezifischer Begriffe in Schule, Hochschule, Aus- und Weiterbildung*. Afra, Frankfurt am Main.

Mandle H. and Fischer F. (2000) Wissen sichtbar machen. *Wissenmanagement mit Mapping-Techniken*. Hogrefe, Göttingen.

Mitra S. and Rana V. (2001) Children and the Internet: Experiments with minimally invasive education in India. *British Journal of educational technology* 32, No. 2, pp. 221-232.

Rumelart D. E., Lindsay P. and Norman D. A. (1972) A Process model for Long-term Memory. In *Organization of Memory*, E. Tulving and W. Donaldsen (eds.), Academic Press, New York.

Winograd T. and Flores F. (1986) *Understanding Computers and Cognition. A new Foundation for Design*. Ablex Publishing Corporation, Norwood, New Jersey.

Let's teach informatics - empowering pupils, students and teachers

Ludger Humbert
Didactics of Informatics, University of Dortmund, Otto-Hahn Strasse, D-44221 Dortmund, Germany

Humbert@ls12.cs.uni-dortmund.de

Key words: Basic Concepts, Classroom Teaching/Practice, Didactics of Informatics, Evaluation, Research

Abstract: Our goal is to develop a didactical structure for a compulsory informatics course. This structure depends on certain fundamental elements, and these will remain in the long run. Results from curriculum research on informatics in secondary education are used to decide which aspects must be included in the subject teaching of informatics. After describing the didactical structure the paper focuses on how successful teaching and learning processes should be organised to enable students to take an active role in long-term learning, express their requirements as users of modern technology; and learn skills and acquire competences in using informatics systems. We propose a modular course to fulfil these requirements. This has been set up to enable some modules to be taught at the secondary I-level (K5-10) and others at secondary II-level (K11-13). These modules enable schools to determine the manner in which informatics will be taught. Our research findings are based on evaluation of informatics as a compulsory course for 6th graders at a German Gymnasium, evaluation of a curriculum for 11th graders at a comprehensive school, and a survey of teachers' opinions on teaching informatics. The paper concludes by stating how the project will move forward and identifies the key questions that will be addressed.

1. WHAT TO TEACH IN INFORMATICS

Only a few schools in Germany have included informatics as a compulsory course for all pupils. However, in the next few years, more schools will establish a compulsory course as all pupils need certain basic knowledge. It is too time consuming to prepare pupils to use specialist applications without giving them a basic knowledge of informatics on which to build in other subjects.

From the point of view of competences and skills the basics are twofold:

- Pupils have to be educated in the principles which enable the process of extracting information from data and those which allow the formulation of information so that it can be represented in data in order that informatics systems can process, transfer and convert it.
- In addition to knowing how the processes are carried out theoretically in an informatics system pupils should receive practical experience in acting with systems so they use theoretical elements in a practically oriented way.

Both approaches to the subject are interwoven and all pupils should acquire basic elements in informatics to support practical experiences in the field of handling information systems. If pupils are educated in informatics they will be able to transfer their knowledge from one application field to another with the help of cognitive models.

The outcome for students in Germany has been described thus:

"... most German students' information literacy is insufficient. German students consider the supply of digital information to be confusing. They feel unable to evaluate the results of their search for information. However, the Internet is the most frequently used medium in searching for scientific information although the use of this medium rarely follows systematic rules. Many German students confine themselves to simply 'browsing the net'. German universities have not taught them how to use the new media systematically. That is why students had to acquire this knowledge as autodidacts." (Klatt et al 2001)

It is important that informatics systems are used (not only at the school level) in conjunction with a field of application. If we examine concrete curricular elements it is apparent that they are currently restricted to some major applications such as word processing, calculation programs, database programs and presentation software. In most cases students only learn how to handle concrete applications. This is not educational informatics - it is a surrogate for handling situations and can, in no way, replace the teaching of informatics. It's comparable to other subjects taught at school: "schools don't teach boiler design - they teach thermodynamics." (Shaw 1992)

We carried out research to describe and to prove various elements of teaching informatics as a subject at school and we wrote proposals to:

- establish a connection between the various elements in teaching informatics as a subject at school (Hubwieser et al 1997);
- prove the hypothesis that it is possible to decide which elements of teaching informatics are fundamental (Schwill 1994).

Informatics should be taught as a subject, developing and integrating correct models of informatics systems in teaching processes. Pupils and students learn best with concrete examples. However, these applications are based on informatics models and, to understand how these products work, knowledge of informatics is imperative.

Over time it has become clear that the object-oriented model is capable of supporting the process of setting up an informatics system as it uses one language throughout. This process works the other way round as well. One is able to "re-engineer" the process for educational purposes. In analysing, deconstructing or describing the functionality of some well known applications in this way pupils are able to build up their own mental model. This enables pupils to criticise the concrete realisation - a first and mandatory step in developing the ability to look at applications with a background in informatics.

1.1 Conceptual elements

We must remember that working on informatics systems is now no longer associated with isolated "personal computing systems" but rather takes place in networked environments. Every introduction must take into account the fact that pupils/students should know the basic elements "behind the scenes" before they go to work in such a complex environment. This is necessary to protect their personality, which should not be damaged by the learning processes. Basic knowledge regarding Netiquette, based on network protocol structures, must be understood.

Another element concerns the basis for constructing and analysing (deconstructing) solutions for problem areas in an object-oriented manner. Setting up a model normally works in a constructive manner, but some research has been done to find ways of decomposing object-oriented software systems (Schulte 2001).

We should not rely on one paradigm - students should be aware that it is possible to work with several paradigms to solve problems. One of the major decisions in the problem solving process is how to find the appropriate paradigm for modelling special elements in the specific problem field.

1.2 Structuring - modularising

To structure teaching processes it is essential to look at cognitive adoption. The process of building up knowledge and competence in informatics must be designed to enable pupils not only to develop skills in handling informatics systems. They must also be able to discern structures and develop mental models of informatics systems with which they can work successfully:

"Because the details of the technology change from day to day, keeping up with those details is difficult and often unproductive. Therefore the study of the subject must concentrate on the fundamental scientific principles and concepts of the field." (ACM 1997)

Modern curricula are structured in a spiral and a project-oriented manner. This segmentation leads to entities which are taught at several cognitive levels - at first pupils learn more elements associated with concrete applications and later on the scope of the first model is restructured on higher cognitive levels. Curricular elements in informatics should structure the field of informatics to create a subject-oriented terminology.

This would help when:

- working on concrete applications;
- discussing the use of terms in informatics;
- it is used in all the courses in informatics;
- working on informatics in other subject fields.

It is also essential that theoretical elements of informatics should be taught at every cognitive level. These elements cannot be ignored due to the lack of time. The following modules have been identified - network and distributed processes; modelling informatics systems; elements of the theory of informatics; applications in specific fields (linguistics, mathematics) and their background based on informatics. Because of the central role of the first element, network and distributed processes must underpin the curriculum. The other elements should be mixed and no one theme should span more than a quarter of the school year.

2. HOW TO TEACH INFORMATICS

2.1 Using models

Nearly all types of models that have ever been developed are used in informatics (Thomas 2001) so the models which have to be taught must be carefully selected. Building upon an abstract layer of structures to implement actions on data is not easy but there is evidence for finding problem fields where students can find ideas for formulating problems and getting ideas for solutions. This serves as a starting point in arriving at successful abstraction and then putting it into practice.

It is important to determine which elements must be taught to all pupils to enable them to integrate these elements in their own subjects and establish a basic knowledge for their future. In addition it is necessary to integrate an historical perspective as well as social elements in teaching informatics. In this way students learn how and why certain rules have been introduced. For example "to log in" should imply that a student not only can log-in but also knows what takes place behind the scenes. Talking about Netiquette offers a way into discussion on how connectionless network protocols provide the opportunity for forging the sender's e-mail address.

On the other hand it is imperative that students have a deeper insight of the process of setting up the deconstruction of a real application so they can see how informatics systems are structured and what the commonalities between the different versions are.

It is essential that every learning process should incorporate historical elements. This allows the developing process to be studied from different points of time or view, enabling the students to identify where critical decisions took place and how the future might be affected. Tom van Weert (2001) has pointed out that:

"Co-operative ICT-supported learning is possible at a higher level"

but we think that it is worth thinking of problem-oriented teaching of informatics at all levels - not only at secondary school or university levels.

Finding a way to structure the complex problem-solving process enables students not only to work on informatics problems successfully, but also to structure the problem fields as well. Students pointed this out when asked what was the effect of being taught informatics for about half a year. A year later even more students felt that they found their knowledge in informatics had been a strategic element in managing problems in other subject areas.

In February 2001, the author conducted research in some Bavarian schools. The aim was to introduce informatics as a subject at the secondary I level at the Gymnasium for 6th grade. It was found that these pupils were capable of learning and using the abstract terms associated with the informatics systems they were using. This was observed in situations where the pupils had to transfer elements from one application context to another. They were capable of doing so because the model they had to describe the application with was a sophisticated one. The description had to be done in an object-oriented manner (Frey et al 2001). This was because they were able to use common terms for different applications and application levels. It is important to realise that such elements are not widely taught!

2.2 Teachers think about teaching informatics

Following a review of the teaching process and the structures associated with teaching informatics the next research element was to interview informatics teachers on the teaching of their subject. We carried out narrative interviews with 16 teachers of informatics. We have transcribed these to find out if there are common elements and where the main differences in vision and practice of teaching informatics exist.

Our first analysis indicates there are some key issues all teachers agree on. They want to use sophisticated teaching software in the classroom. They want such software to be available under a free licence (like GPL) so they are able to adapt it for their learning groups. They are looking for educational software they can not only use but also can modify for specific needs in the adapted teaching and learning environment they want to prepare for their students. Such software gives control back to teachers, so they can decide what concrete elements are visible, what patterns are used in a special user interface and so on.

Nearly all of the teachers who were interviewed told us they are unhappy to leave out the more hardware oriented elements from the teaching and learning process. They told us that the role of informatics in school has changed over the years. It will change even faster in future years because more and more parents want the school to teach children about informatics so they might have the opportunity to play an active role in the future.

3. CONCLUSION - REMAINING QUESTIONS

The results of our research clearly indicate the necessity of finding ways to incorporate informatics as a compulsory course for all pupils at secondary I level.

We must stress that the teaching of informatics must not simply cover the use of application no more than the teaching of physics enables pupils to cable houses or set up power stations. It should provide practical experiences in order that concrete elements, which pupils understand at a theoretical level, are put into practice.

We should not forget the gender breakdown in secondary II courses. Given the choice students who choose informatics as a course at the secondary II level at German gymnasiums in North Rhine Westfalia were male. We have to construct the compulsory course so that, in the future, girls become more involved in informatics.

REFERENCES

Association for Computing Machinery (1997) *Model high school computer science curriculum.*
[http://www.acm.org/education/hscur/index.html]

Frey, E., Hubwieser, P., Humbert, L., Schubert, S. and Voss, S. (2001) *Erste Ergebnisse aus dem Informatik-Anfangsunterricht in den bayerischen Schulversuchen.* LOG IN, 21 (1). pp. 25-37.
[http://ddi.cs.uni-dortmund.de/ddi_bib/forschung/pub/Informatik-Anfangsunterricht.pdf]

Hubwieser,P., Broy, M. and Brauer, W. (1997) A new approach to teaching technologies: shifting emphasis from technology to information. In *Information Technology - Supporting change through teacher education*, D. Passey and B. Samways (eds.), Chapman and Hall, London, pp. 115-121.

Klatt, R., Gavrilidis, K., Kleinsimlinghaus, K. and Feldmann, M. (2001) *Barriers in using digital scientific information at German universities and other higher education institutions - how to develop potentials in academic education.* Lecture at 23rd. DGI-Online Conference 2001, Frankfurt am Main.
[http://www.stefi.de/download/english.pdf]

Schwill, A. (1994) Fundmental ideas of computer science. *EATCS-Bulletin*, (53): pp.274-295.
[http://www.didaktik.cs.uni-potsdam.de/Forschung/Schriften/EATCS.pdf]

Schulte, C. (2001) Changing code - notes on the interdependency between informatics and society. In *Book of Abstracts - Networking the Learner*, 7th. World Conference on Computers in Education (WCCE 2001), J. Andersen and C. Mohr (eds), page 167, Gyldendal Education, Copenhagen.

Shaw, M. (1992) We can teach software better. *Computing Research News*, 4th. September, pp. 2, 3, 4, 12. Reprinted in *Journal of Computer Science Education*, 7, 3, Spring 1993, pp. 4-7.
[http://spoke.compose.cs.cmu.edu/shaweb/edparts/crn.htm]

Thomas, M. (2001) Vielfalt der Modelle in der Informatik. In *Die Informatik und Schule - Informatikunterrich und Medienbildung* INFOS 2001, R. Keil-Slawik and J. Magenheim (eds.), Springer, Berlin. pp. 173-186.

van Weert, T. J. (2001) Cooperative ICT-supported learning - A practical approach to design. In *Die Informatik und Schule - Informatikunterrich und Medienbildung* INFOS 2001, R. Keil-Slawik and J. Magenheim (eds.), Springer, Berlin. pp. pp. 47-61.

Key Decisions in Adopting Algorithm Animation for Teaching

Guido Rößling
FG Telecooperation, FB 20, Darmstadt University of Technology, Alexanderstrasse 6, D-64283 Darmstadt, Germany

roessling@informatik.tu-darmstadt.de

Key words: Algorithms, Distance Learning, Education, e-learning, Learning Materials

Abstract: Algorithm Animation is becoming increasingly popular with several educators. However, certain key questions have to be addressed before an informed decision on a given system can be made. This paper identifies several such questions and provides a brief overview of some algorithm animation systems. The key decisions are grouped in four categories: system type and animation generation approach; display properties; import and export facilities; and didactical requirements. After considering these requirements, selecting a particular system should be easier for educators.

1. INTRODUCTION

Algorithm Animation (AA) has received increasing interest over the last few years within the education community. This is reflected in the number of publications at educational conferences (such as the yearly ACM SIGCSE and ITiCSE conferences), the establishment of an international Program Visualization Workshop in 2000 and a Dagstuhl Seminar in May 2001. However, many educators still wonder what algorithm animation is and how it could help them in their teaching.

Simply put, algorithm animation is the dynamic visualization of the processes that make up algorithms and data structures. Thus, any tool that illustrates the execution of a given algorithm can be counted as AA. Today many different systems exist. Educators interested in employing AA therefore have to make an informed decision to adopt a given system, or decide to develop their own. However, there are vast differences between the available tools. In this paper, we will examine typical examples of the different approaches taken by a restricted selection of AA systems. We hope that the survey is helpful as a guideline for deciding on a given system. Ideally, it will also help generate interest in AA from educators who are still unsure whether or not AA may assist them in their teaching.

The basic reference for AA is the book "Software Visualization" edited by John Stasko et al (1998a). This book includes two elaborate taxonomies for classifying AA systems. However, these taxonomies do not focus on the factors that influence the adoption of a particular AA system for teaching. This paper focuses on outlining central aspects that should be taken into account when deciding whether or not AA should be adopted into teaching.

2. KEY DECISIONS

We can classify the key decisions that must be made into five different areas - the type of system, how animations are generated or edited, content display properties, import and export facilities, and didactical requirements. A complete discussion of all the requirements that can be placed on AA systems is beyond the scope of this paper - it focuses instead on outlining the central issues that need to be addressed.

2.1 AA System Type and Content Generation

The type of system represents the basic approach followed by a given system. We regard four different user roles - *programmer, developer, visualiser* and *user* (Price et al 1998). The *programmer* role implements a given algorithm without regard for animation purposes. In fact, the programmer may be unaware of any plans for future animation of the code. The *developer* role covers the design and construction of the AA system used to portray the animation content. Content generation and editing is the task of the *visualiser* role. Finally, the *user* role represents the end-user who studies the animation and possibly interacts with it.

The AA system can be classified as belonging to one of six different types according to the content generation method employed. *Topic-specific* systems are implemented for a certain target area, for example graph algorithms. They may provide very good support for this field of expertise, but are often not usable outside their focus. We examined the applications

RLE, *Quadtree* and *JPEG* which were representative of this category of systems. They portray the properties of the image compression and encoding approaches identified by Khuri and Hsu (2000). The packages provide verbose explanations for the topic area but each is "hard-wired" to a particular problem and thus cannot easily be reused. The content is generated from the user input, so that the visualiser role is not actually able to generate an animation according to his or her preferences.

Systems that allow content generation within a Graphical User Interface (*GUI*) give the greatest degree of design freedom to the visualiser. There are two main drawbacks. First, all of the content must be generated manually which is very time-consuming. Secondly, as the content is typically generated independently of the algorithm that is to be illustrated, the visualiser is responsible for ensuring that the algorithm is portrayed correctly. On the other hand, manual generation also allows abstract displays that are based on pseudo-code, use metaphors, or focus only on certain aspects of the algorithm. The selection of AA systems that incorporate visual generation is rather limited and, apart from presentation tools such as PowerPoint and Star Impress, the main system currently in use is ANIMAL (Rößling and Freisleben 2000).

API-based generation uses method invocations from a special library for visualizing the content of the algorithm. The necessary method invocations are usually embedded in a modified version of the underlying algorithm. A new animation is therefore generated simply by restarting the algorithm. One of the best known of the older animation systems, *XTANGO*, employed a special C-based API (Stasko 1998b). API-based generation can be very elegant. However, this elegance depends on the features offered by the API. The method invocations embedded in the source code modify the running time and may also result in side effects. The visualiser has to be skilled in the programming language underlying the API, and is typically restricted to algorithms implemented in the same underlying programming language. Additionally, the animation is usually performed immediately, and there may be no provision for exporting to a file so that the animation could be reused.

Scripting-based generation parses a set of commands to generate the animation content. Typically, the commands are gathered in a file, in this case allowing reuse of the animation content. As most scripting languages are relatively straightforward in their notation the visualiser can also modify the commands using any text editor. The commands can be generated manually by simply typing them in an editor or can be generated from the algorithm, typically by sending the appropriate commands to the standard output or error terminal and redirecting this output to a file. Two typical representatives of scripting-based systems are *JAWAA* (Pierson and Rodger 1998) and *JSamba* (Stasko 1998b). The scripting languages employed by the systems are similar but incompatible. While comparable in their general approach, there are interesting differences between the systems. For example, JAWAA is specifically geared for Computer Science content, providing special commands for common data structures and operations thereon, such as trees or stacks. JSamba is more general, requiring the user to assemble stacks manually from base components. However, JSamba provides a simple zooming function that can be used by the visualiser to focus the user's attention.

Declarative generations embed special logical predicates into the algorithm, usually within comments. The comments are extracted and evaluated by special preprocessors or full-fledged virtual machines. Alternatively, a modified compiler can be used that replaces the predicates with appropriate pieces of code. Leonardo (Crescenzi et al 2000) is one example system that follows declarative visualization.

Finally, some systems use *code interpretation* to visualise the algorithm. One of the best known systems is *ZStep 95* (Liebermann and Fry 1998). It evaluates and visualises LISP code. It is interesting to note that no visualiser interaction is required for generating the animation, as this is performed automatically. This also means that visualisers cannot tailor the appearance of the animation to their preferences.

In general, there is no single "best" method of animation generation and editing - it really depends on the preferences and skills of the visualiser. *Manual generation* offers the greatest flexibility of presenting the content - however, it is easily the most time-intensive method. Direct *interpretation* of the underlying code causes the least work for the visualiser - however, the appearance can usually not be adjusted. *API-based* and *declarative* generations are "cleaner" ways for embedding visualization generation statements into the underlying algorithm than is the case for *scripting*; however, the visualiser needs a certain amount of familiarity with the underlying programming language and logic to use the first two techniques. Finally, *topic-specific* systems are often highly specialized, but may restrict the freedom of the visualiser with regard to adjustment of the display. Additionally, they are usually ill suited for use outside their specific area of experience. Alas, most systems do not combine more than a single generation approach, with the exception of ANIMAL (Rößling and Friesleben 2000). It offers GUI, scripting and API-based animation generation.

2.2 Animation Display Criteria

To address different needs of the user, the animation display should address several issues in its display front-end. First of all, the animation canvas should be resizable to allow it to fit a smaller or larger display (a notebook instead of a desktop PC). For presentations in larger lecture halls, the content should also be freely scalable. Of the systems discussed here, only ANIMAL and JSamba offer zooming - the former within the display GUI, the latter only by animation commands. Thus, the presenter of an animation cannot easily adapt the zoom scale to one appropriate for presentation using a beamer.

The second issue regards the controls for steering the behaviour of the animation display. Ideally, the full functionality of a video player should be provided - including fast forward and reverse playing. Currently, the number of systems that offer "real" reverse playing is very limited, due to the difficulty in implementing free reverse playing. In 2001 researchers stated that efficient reverse animation playing and rewinding can be considered one of the foremost "open questions" in AA (Anderson and Naps 2001). Two representative tools that offer this functionality are ANIMAL and ZStep 95 (Liebermann and Fry 1998). Most other systems are restricted to offering *play, pause* and a *step forward* functionality that shows the next state without animating the content.

The visualiser should also be able to specify how a given animation step is reached from its predecessor. In some systems, such as JAWAA and JSamba, the next step is automatically executed after the current step has completed execution. Optionally, the visualiser may issue an appropriate command to pause the display. The image compression packages by Khuri and Hsu (2000) show the next state only after a user interaction. ANIMAL combines both approaches: for each animation step the visualiser can define if the next step is shown automatically after a certain delay or only after a user interaction. In addition to the *pause* button offered by many systems, ANIMAL also offers a "slide show" mode. This links all steps with a predefined delay. The mode is automatically left when the pause button is depressed.

2.3 Animation Import and Export

Most current animation systems offer no support for exporting the animation content to a different format or importing foreign formats. Some systems, such as the code interpretation-based WinHIPE (Velázquez-Iturbide and Presa-Vásquez 1999) offer at least an export to graphic formats such as bitmaps or GIF images. The lack of import and export facilities in most systems is surprising, as the provision of these facilities is usually not very difficult, and provides powerful incentives for users to adopt the system due to the possibility of content reuse. ANIMAL again is one of the exceptions to this rule, offering export to various image formats such as JPEG and PNG, XML streams, and Quicktime videos. Some of the export formats require special libraries that have to be installed separately. Therefore, the export formats are treated as extensions and not provided in the basic distribution.

2.4 Didactical requirements

Several didactical requirements for improved retention have been discussed recently in the AA community. One of the foremost of these is interactive prediction, as incorporated in the *JHAVÉ* system (Naps 2001). Here, the user is asked at fixed points to predict what the next operation shown by the AA system should be. In *JHAVÉ*, the prediction is performed as a multiple-choice quiz. Some other systems such as *PILOT* (Bridgeman 2000) let the user actually "perform" the operation by interacting with the display. This type of interaction requires a great deal of context awareness on the part of the system, and is therefore usually limited to topic-specific systems. The main benefit of interactive prediction, and generally of AA, is the raised interest and motivation of the students. Recent studies disagree on the learning effect inferred by prediction or AA in general. Obviously the learning effect is also largely dependent on the way the content is presented and the way that users can interact with it.

Naps argues for the value of incorporating links to external documentation for the displayed content (Naps 2001). The documentation is typically encoded in HTML and should provide additional information about the current display and the algorithm in general. Naps differentiates between fixed or "static" documentation, *algorithm-sensitive* documentation that is aware of the current stage of the displayed content, and *dynamic* documentation which is algorithm-sensitive and also incorporates the concrete values of the current animation display.

3. CONCLUSION

Many educators are willing to employ algorithm animation in their lectures, but are not sure how to do it. One large part of this uncertainty is the lack of knowledge about the available systems and how they can be compared and evaluated. It is difficult to obtain a full list of available animation systems as their number was estimated at 150 in 1998 and has risen since then (Price et al 1998). However, there are some good resources which give information about AA and AA systems:

- http://www.cs.hope.edu/~alganim/ccaa/index.html
- http://www.cs.cofc.edu/~mccauley/edlinks/
- http://www.animal.ahrgr.de

In this paper we have focused on presenting some of the key decisions that educators should consider when they decide to adopt AA for their courses.

The foremost consideration is the approach used for generating new animations. This has a large effect on the system's usability and depends on the educator's skills and preferences. Other issues relating to the display of the animation content, import and export, and didactical requirements have been discussed. The breadth of the topic is so wide that several issues could not be covered in this paper. We trust that our analysis will prompt readers to think deeply about the requirements for, and of AA systems.

REFERENCES

Anderson, J. M. and Naps, T. (2001) A Context for the Assessment of Algorithm Visualization Systems as Pedagogical Tools. In *Proceedings of the First International Program Visualization Workshop*, 2000. E. Sutinen et al (eds.), University of Joensuu Press, Joensuu, Finland. pp. 121-130.

Bridgeman, S. (2000) PILOT: An Interactive Tool for Learning and Grading. In *Proceedings of the 31st SIGCSE Technical Symposium on Computer Science Education*, 2000. ACM Press, New York. pp. 139-143.

Crescenzi, P., Demestrescu, C., Finocchi, I. and Petreschi, R. (2000) Reversible Execution and Visualisation of Programs witth LEONARDO. *Journal of Visual Languages and Computing*, Vol. 11, No. 2, April, pp.125-150.

Khuri, S. and Hsu, H-C. (2000) Interactive Packages for Learning Image Compression Algorithms. In *Proceedings of the 5th Annual Conference on Innovation and Technology in Computer Science Education* (ITiCSE 2000), J. Tarhio (ed.), ACM Press, New York. pp. 73-76.

Liebermann, H and Fry, C. (1998) ZStep 95: A Reversible, Animated Source Code Stepper. In *Software Visualization – Programming as a Multimedia Experience.* J. Stasko et al (eds.), MIT Press. pp. 277-292.

Naps, T. (2001) Incorporating Algorithm Visualization into Educational Theory: A Challenge for the Future. *Informatik / Informatique Special Issue on Visualization of Software*, April 2001. pp. 17-21.

Price, B. et al (1998) An Introduction to Software Visualization. In *Software Visualization – Programming as a Multimedia Experience.* J. Stasko et al (eds.), MIT Press. pp. 3-27.

Pierson, W. and Rodger, S. (1998) Web-based Animation of Data Structures Using JAWAA. In *Proceedings of the 29th SIGCSE Technical Symposium on Computer Science Education.* J. Lewis (ed.), ACM Press, New York. pp. 267-271.

Rößling, G. and Freisleben, B. (2000) The Animal Algorithm Animation Tool. In *Proceedings of the 5th Annual Conference on Innovation and Technology in Computer Science Education* (ITiCSE 2000), J. Tarhio, (ed.), ACM Press, New York. pp. 37-40.

Stasko, J. et al (1998a) *Software Visualization – Programming as a Multimedia Experience.* MIT Press.

Stasko, J. (1998b) Smooth Continuous Animation for Portraying Algorithms and Processes. In *Software Visualization – Programming as a Multimedia Experience.* J. Stasko et al (eds.), MIT Press. pp. 103-118.

Velázquez-Iturbide, J. Á. And Presa-Vázquez, A. (1999) Customization of Visualizations in a Functional Programming Environment. *Proceedings of the 29th ASEE/IEEE Frontiers in Education Conference*, IEEE Press, New York, pp. 12b3 22-28.

Design Pattern - a topic of the new mandatory subject informatics

Markus Schneider
Department of Informatics, Technische Universität München, 80290 München, Germany

markus.schneider@in.tum.de

Key words: Curriculum Changes, Didactics of Informatics, Informatics, Teacher Education

Abstract: Design patterns are commonly used in object-oriented modelling for the solution of frequently appearing design problems. Until recently their use was restricted to the domain of software engineering, However, object-oriented modelling is an important topic of the new mandatory subject informatics, which will be introduced during 2003 in all the four hundred Gymnasiums of Bavaria. In this paper we want to show that design pattern might add a valuable contribution to such a mandatory subject, which has its main focus on teaching the substantial concepts of informatics and on stimulating mental models to give real understanding of informatics.

1. INTRODUCTION

The designer of object-oriented software systems has to deal with recurring design problems such as modelling a recursive data structure, designing the menu of an application or modelling the interaction between the user interface and the core of the application. For such problems there are design patterns that provide a general solution. These solutions are visualized graphically using the Unified Modelling Language (UML) (Booch et al 1997).

The first systematic collection and classification of these design patterns was presented in 1995 and a year later another system of patterns was published that used a different system of classification (Gamma et al 1995; Buschmann et al 1996). Neither provided a formal definition of the concept "design pattern". However Gamma described the concept informally as:

"The design pattern ... are descriptions of communicating objects and classes that are customised to solve a general design problem in a particular context.";

and Buschmann as:

"A pattern addresses a recurring design problem that arises in specific design situations and presents a solution to it."

Until recently the use of design patterns was restricted to the domain of software engineering. Their use in informatics at school wasn't considered. Currently, there are efforts to introduce the basic concepts of object-oriented modelling in the proposed mandatory subject of informatics at the Bavarian Gymnasium (Hubwieser 2002). To this end, standard business applications are analysed and the structure of these applications is modelled by object-oriented means as early as the 6th grade. The main idea is to familiarise the students with the basic structures of informatics that underlie an application and stimulate the development of cognitive structure to serve as mental models for real understanding. We wish to investigate whether the development of these cognitive structures could be supported by design patterns, since they represent graphically many of the basic concepts discussed in the proposed mandatory subject of informatics.

In the second part of this paper we show that the composite pattern is a frequently appearing design pattern in the projected curriculum of the 6th grade. Furthermore, we illustrate the didactic function of this pattern. Then we point out the extent to which the observer pattern describes substantial aspects of interactive applications and we show the usefulness of utilising this pattern in school. Finally we discuss general criteria to select design patterns. These could be useful at school level. We also present many of the unresolved questions on the subject of design patterns.

2. THE COMPOSITE PATTERN AND ITS APPEARANCE IN THE SUBJECT OF INFORMATICS

In this section we will show that the composite pattern already appears in the curriculum of the 6th grade. However, as this pattern contains elements of object-oriented modelling, which is not discussed at this level, the composite pattern will not become a key topic of 6th grade study. Nevertheless, it is an important element of background information for the teacher and it is a structure well worth discussing in grades 9 - 13. The pattern provides a possible approach to establishing the basic concepts of informatics.

2.1 Curriculum areas related to the composite pattern

Objects in graphics

In the 6th grade, objects in vector graphical images are the first structures to be analysed and modelled.

For the first time the students have to familiarise themselves with the basic concepts of object-oriented modelling. In the discussion of these simple objects the composite pattern occurs. To illustrate this, let us analyse an image typically used in this phase of education (Hubwieser 1998).

The plan of, for example, an apartment is composed of graphics that represent the various rooms (Figure 1). The graphics of each room consist of graphics that represent the various furnishings. Illustrating these relations graphically a tree-like object diagram can be constructed (Figure 2). For reasons of clarity not all attributes have been specified. In a typical class diagram some attributes are declared (Figure 3).

In this example the composite pattern consists of three classes - Graphics, Geometrical Object and Graphical Object. Thereby, it is essential that every graphic is an aggregate of a certain number of graphical objects and, in particular, a certain number of other graphics. Instantiating this pattern, we get the tree-like object diagram shown in Figure 2.

Abstracting from this concrete example the composite pattern takes the form illustrated in Figure 4. It is worth mentioning that the superclass is built by the Component whereas the Compositum, which represents the system, is a lower class.

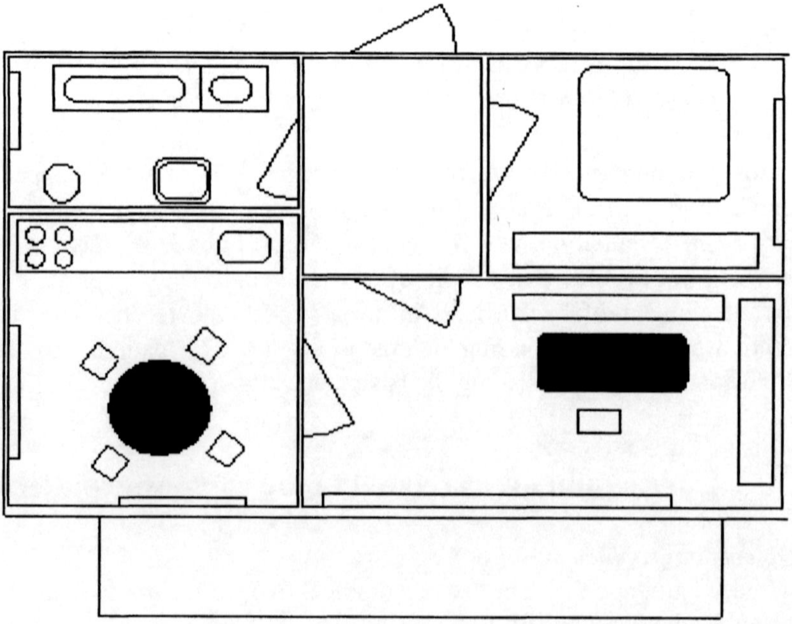

Figure 1. Plan of an apartment

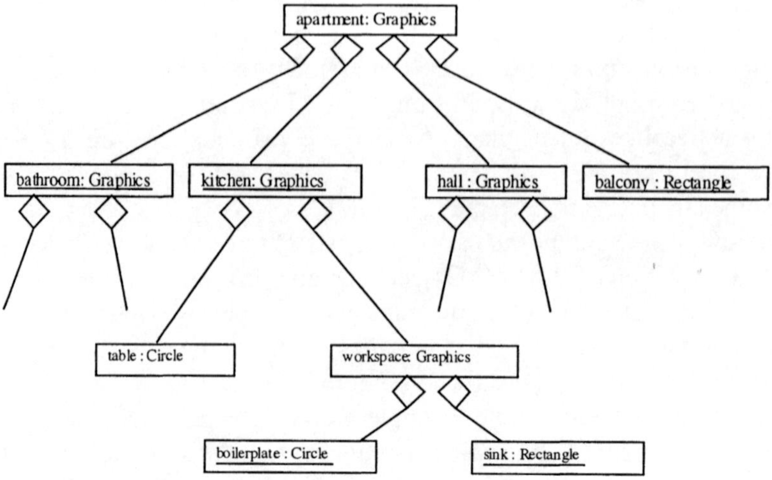

Figure 2. Object diagram of the apartment

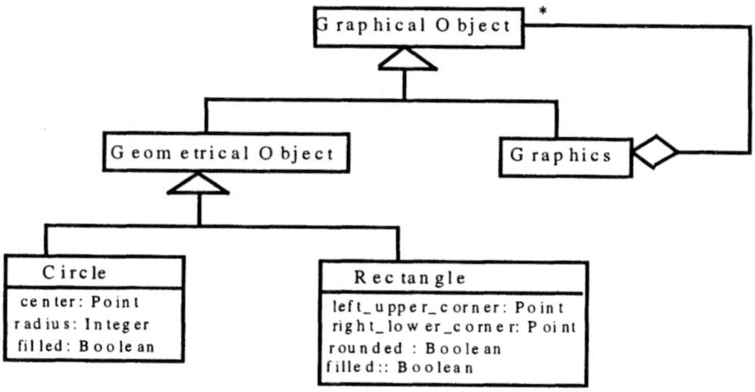

Figure 3. Class diagram of graphics

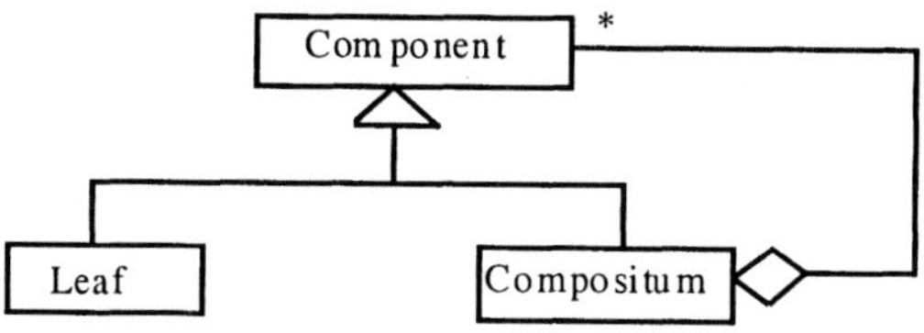

Figure 4. The composite pattern

Thus we have described the concrete scenario, the object diagram, the class diagram and finally the abstract pattern. In the following we will briefly outline the concrete scenario and then present the respective class diagram.

Relations between textual objects

In the 6th grade curriculum the discussion of graphical objects is followed by the object-oriented analysis of textual objects. We also have tree-like object structures for textual analysis. Consider the following scenario. A textual object consists of several sections. These sections are nested or divided into several main sections, e.g. this text is divided into - "Introduction", "The composite pattern" and so on. Then every main section is subdivided and this process of subdivision continues until a single character is reached.

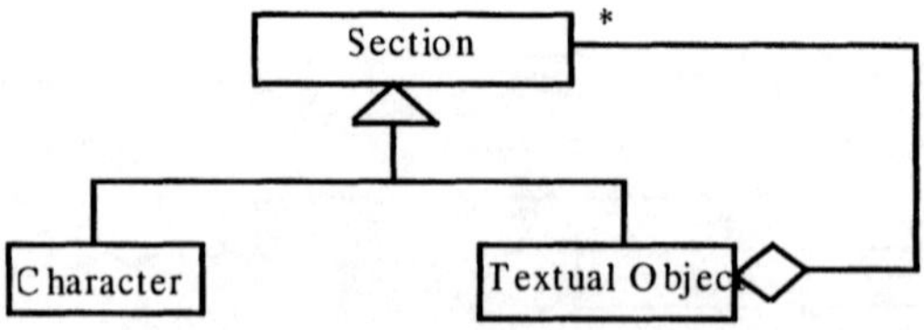

Figure 5. Composite pattern for textual objects

This description suggests the following model - every textual object consists of several sections and every section in turn is either a character or a textual object that aggregates a number of sections (Figure 5).

Hierarchical structures of files and directories

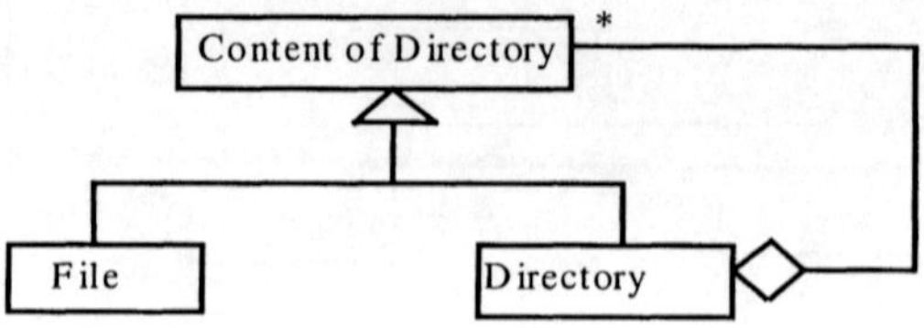

Figure 6. Composite pattern for directory structures

This topic is selected to illustrate the use of the composite pattern. A directory aggregates a certain number of objects. These represent the content of the directory. This content is either a file or a directory and we therefore get the pattern shown in Figure 6.

2.2 Some conclusions

The examples illustrate that the composite pattern is a substantial structure. It occurs at many points in the proposed curriculum for the 6th grade. However, the complexity of this design pattern forbids its direct use as a 6th grade topic as this would require a profound understanding of modelling techniques like inheritance.

Currently, in grades 9 -11, tuition is focused on the improvement of object-oriented modelling techniques established in the 6th grade. In grades 12 and 13 recursive data structures and recursive function calls are discussed

(Hubwieser 2001). We consider that the composite pattern might offer an important contribution to the teaching of such content.

Improvement of the object-oriented modelling skills:

After the introduction of inheritance, the student should acquire an understanding of the interaction between aggregation and inheritance through exploration of practical examples. Obviously familiar examples from the 6th grade could be selected and appropriate class diagrams could be developed in the manner outlined above. This approach offers further advantage as the students would learn (in a very impressive manner) the difference between the instance of a class and the class itself. The tree-like object diagram emerges from the respective composite pattern by instantiation.

Propaedeutics of the topic recursion

Recursion is probably one of the most difficult topics to teach in informatics. Currently recursion is a topic in the 11-13 grades. One possible strategy might be to start with the composite pattern, then discuss recursive data structures and then introduce the recursive function call. This strategy reverses the common approach (Broy 1998) and was proposed by Hubwieser (2000). It has the advantage of starting from known structures - namely the tree-like object diagrams.

3. THE OBSERVER PATTERN

The composite pattern, illustrated in the previous section, is a design pattern that describes, for example, the structure of a text document. The observer pattern, in contrast, is a design pattern for modelling the interaction between certain objects, e.g. the interaction between the document, the user interface and other objects like the clipboard. Essentially, the observer pattern provides two services. On the one hand it decouples the user interface from the methods of the document and on the other hand it allows the design of user interfaces that provide (depending on the state of the document) the functionality that is currently applicable.

Prior to an illustration of the observer pattern the following points should be mentioned. The observer pattern is a pattern that describes substantial structures of interactive applications and therefore it is important for inclusion in 6th grade. Since the complexity of this pattern clearly exceeds that of the composite pattern, the following serves primarily as background information for the teacher of the 6th grade. Nevertheless, it could be used in grades 9-11 in the context of a detailed discussion of object-oriented modelling techniques.

3.1 Introductory example

Consider a typical part of an office application. The user interface contains the menu items "insert", "copy" and "exit". These menu items interact with the document and the clipboard in the following manner:

- The menu item "copy" is active only when a textual object is marked in the document. When performing a mouse click on this menu item the content will be copied into the clipboard.
- The menu item "insert" is active only if the clipboard is not empty. This menu item allows the insertion of the contents of the clipboard into the document.
- When activating the menu item "exit" the content of all open documents should be saved. The objects of the clipboard may be saved for later use.

Now, we can model the relationships between the objects "Document", "Clipboard", "Copy", "Insert" and "Exit". The elements of the user interface should provide event-handling methods. These should use appropriate methods of the document or clipboard. In this way the user interface can be separated from the functionality of the document or the clipboard.

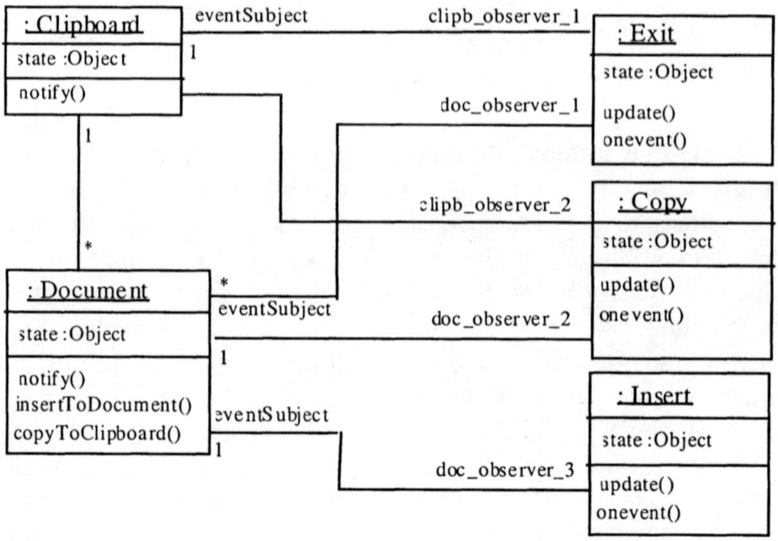

Figure 7. Object diagram of a subsystem of an office application

Now look at the object diagram (Figure 7). It describes the objects and the relationships between them. On the right-hand side the object diagram contains several observers. These elements of the user interface observe the clipboard and the documents. The clipboard and documents are called observables. Depending on the state of the observed object the observers take on a certain state, for example, the object "Copy" is active when a part of the document is marked. The updating of an observer results from the notify method - as soon as the state of an observable changes this method will be called. Then the notify method calls the update methods of all associated observers.

3.2 Abstraction

Abstraction from the above example results in the observer pattern described in Figure 8. It is significant that every call of a method of the observable results in a call of the method notify. This method in turn causes all observers to update themselves.

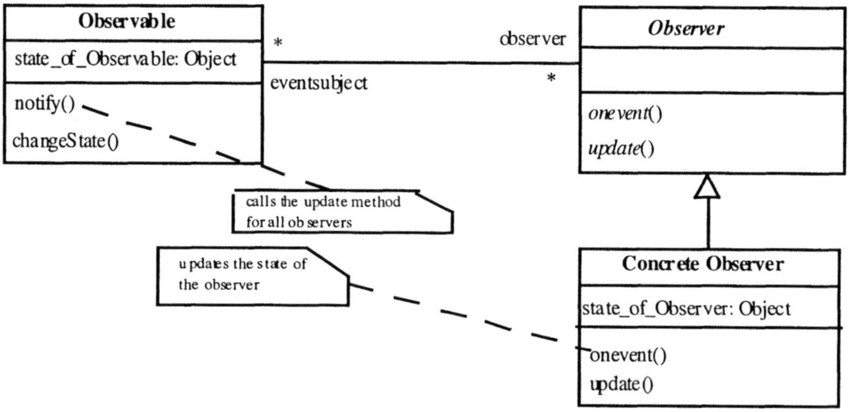

Figure 8. The observer pattern

In the class diagram (Figure 8) we introduced an abstract superclass. For an introductory understanding of the observer pattern it is not necessary to do so, since one might associate the observables with the concrete observers directly. However, for more profound understanding the use of an abstract class is helpful. It reinforces the fact that the event handling method and the update method are properties of all observers.

The observer pattern is a pattern that describes the interaction between objects of the user interface and the application. Therefore, the dynamic behaviour of the involved objects is of particular interest. The sequence diagram (Figure 9) shows the behaviour of the objects "Document", "Clipboard" and "Copy".

First a certain text area is marked in the document. This modification of the documents state activates the notify method. This in turn calls the update method of the observer object "Copy". Now, if the object "Copy" registers an event, the event-handling method of this object calls the method of the document and inserts the marked text area into the clipboard. The clipboard in turn notifies all observers by calling the respective update method.

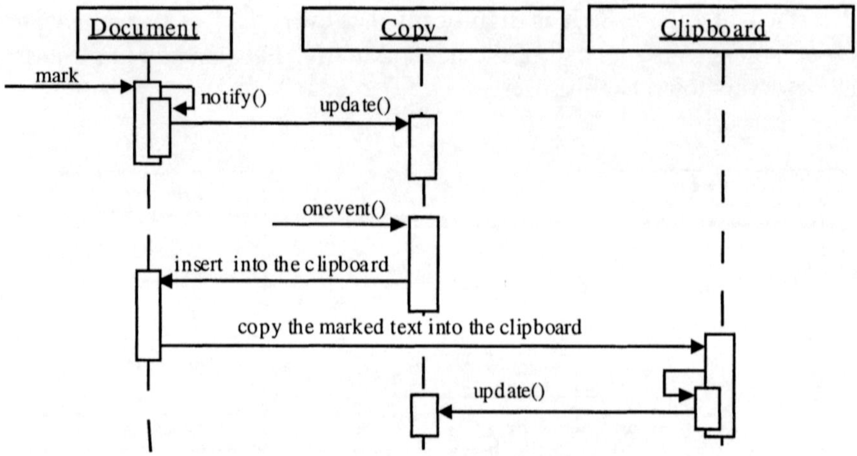

Figure 9. Scenario with observers and observables

3.3 More interesting subsystems of interactive applications with observer pattern features

In the following, we briefly outline how the observer pattern occurs in some subsystems of interactive applications:

- A document has been created. The menu items "Save" and "Save as" remain inactive as long as the user has made the first modifications;
- A document can be closed only when the content of the document has been saved.
- When opening a document the "Ok" button is not activated until a file name has been selected or declared.
- The form of the mouse button varies according to the position.

3.4 More conclusions

The observer pattern is complex and it cannot be used in pure form in the grades 6, 9 and 10. However, in a simplified form, for example without the use of abstract classes or in the form of an object diagram, this pattern might be discussed.

As is apparent from the above illustrations the modelling techniques that are used cause no difficulties. It is more likely that the methods used in the observer pattern could be a problem since substantial features of these methods cannot be modelled graphically.

Separation of the user interface and the core of the application

With regard to object-oriented modelling (proposed for the mandatory subject informatics at the Bavarian Gymnasium) it is essential that the student should be aware that the user interface is simply an interface to the methods of the application and is not the application itself. Once aware of this separation students will be able to understand the meaning of object-oriented modelling and can develop the cognitive structures which lead to a profound understanding of informatics. This concept of separation is described by the observer pattern and this pattern is therefore an ideal tool for stimulating the development of the above mentioned cognitive structures.

4. ARE THERE MORE PATTERNS WITH RELEVANCE FOR THE SUBJECT OF INFORMATICS?

For the composite and observer pattern the relationship with topics of the projected subject informatics is clear. The question arises, which other design patterns might be also of relevance for this subject? There are candidates like the interpreter pattern in the context of formal languages or the state pattern to model the state-dependent behaviour of an object. The more substantial question in this context is - Do we have general criteria to select the patterns whose use might be of value to the subject of informatics?

The following criteria are suggested:

- The pattern has to represent a substantial concept of informatics.
- The pattern has to be of general interest. Special patterns, which are utilised in some specific context only, are normally not suited.
- The pattern has to provide real support to the teaching of the represented concept and therefore provide a didactical function.
- The pattern should not be too complex such that the students would not understand it easily or fully.

- The understanding of the respective pattern should not require too much background information about a specific programming language.

These criteria could be classified into two categories. On the one hand we have a contextual category (criteria 1 and 2) and on the other hand we have a didactical category (criteria 3, 4 and 5). For our purposes a systematic analysis of design pattern has to clarify the following questions:

- How can the design patterns be classified using the categories - contextual, didactical?
- What are the consequences of this classification with respect to the use of the patterns in the subject of informatics?
- How can these patterns be integrated into the subject of informatics?

4.1 Known classification schemes

Both Gamma and Buschmann proposed a system of patterns. These systems are not ideal for the purposes described above.

Gamma et al divided the patterns into structural, behavioural and creational patterns. Creational patterns are patterns which systematise the instantiation process. Since this definitely involves questions on the level of a specific programming language, creational patterns might not be suitable for use in schools.

The other classification proposed by Buschmann et al divides the pattern into architectural pattern, design pattern and idioms. Idioms deal with the details of the implementation and therefore can be excluded for our purposes. The criteria for the classification of the architectural and design patterns are again too coarse and unspecific to estimate their use globally.

5. CONCLUSION

In summary we conclude that the two known classification systems are helpful in pre-selecting appropriate patterns. However, for tuition we need systematic analysis that classifies the patterns in the manner outlined above. The application of such a classification could clarify the questions and issues such as which design patterns ought to be used at school, which basic concepts of informatics are represented by the selected patterns and in which manner will these patterns be introduced in the mandatory subject of informatics?

REFERENCES

Booch, G., Rumbaugh, J. and Jacobson I. (1997) *Unified modelling Language; Semantics and Notation Guide 1.0*; Rational Rose Software Corp., San Jose, California.

Broy, M. (1998) Informatik. *Eine grundlegende Einführung*, Auflage, Springer.

Buschmann, F., Meunier, R., Rohnert, H., Sommerlad, P. and Stal M. (1996) *A system of pattern*. Wiley and Sons.

Gamma, E., Helm, R., Johnson, R. and Vlissides J. (1995) *Design patterns; Elements of Reusable object-oriented Software*. Addison Wesley.

Hubwieser P. (1998) *Didaktik der Informatik*. Springer.

Hubwieser P. (2000) *Rekursion im diaktischen Querschnitt*. Habilitationsvortrag.

Hubwieser P. (2001) Objektorientierte Modellierung von Standardsoftware im Anfangsunterricht Informatik. *INFOS 2001*.

Hubwieser P. (2002) Mental Models enhancing user skills. Unpublished paper.

Learning to solve ICT/informatics-based problems

Mary Webb
Department of Education and Professional Studies, King's College London, Franklin-Wilkins Building, London SE1 9NN, United Kingdom

mary.webb@kcl.ac.uk

Key words: Cognition, ICT, Informatics, Learning Models, Pedagogy

Abstract: The need for developing a pedagogy for ICT/informatics has been identified. While a great deal of research and thinking has focused on theories of learning in relation to ICT this has concentrated on using ICT rather than learning about ICT or Informatics. The links between theories of learning and practice of teaching are, therefore, much less clear for ICT. The nature of the ICT curriculum is examined in order to identify how particular theories of learning may inform the development of pedagogy for ICT. The components of a model of problem solving that fits in with the primary emphasis of ICT/Informatics courses are explored in relation to key learning theories. Important pedagogical implications are identified.

1. INTRODUCTION

Recent evidence of standards of teaching and learning of ICT in England still reveals significant shortcomings. These deficiencies, together with the perceived importance of developing ICT knowledge and skills, has led to calls for the elucidation of a pedagogy for ICT. In England the term ICT is used to describe both the study of computers and other technologies that are used for communications and Information Systems and the use of ICT tools and processes in other areas of the curriculum. This paper focuses on ICT as a subject.

An examination of the pedagogical reasoning process as it applies to the teaching of ICT, revealed a number of problems for ICT teachers and led to the identification of actions to address these problems (Webb, in press). These include classroom-based research, informed by appropriate theories of learning, to identify pedagogical content knowledge and subject specific interpretations of general pedagogical knowledge.

In long-established subjects such as science the links between theories of learning and practice of teaching have developed over many years (Adey 2001). While a great deal of research and thinking has focused on theories of learning in relation to ICT this has concentrated on using ICT rather than learning ICT or Informatics. Therefore the links between theories of learning and practice of teaching are much less clear for ICT.

Before we can analyse in detail how specific theories of learning might inform the development of pedagogy for ICT, we need a clear idea of the nature of the ICT curriculum and the learning processes that are involved. In this paper I identify and discuss the nature of key concepts and processes of ICT and clarify the types of knowledge, skills and processes that students are expected to learn. This facilitates examination of how selected theories of learning may inform the development of an appropriate pedagogy. A first step is the clarification of the nature of the curriculum for learning ICT itself.

2. THE ICT CURRICULUM

In the UK, courses at GCSE level aimed primarily at 14-16 year-olds, develop knowledge, skills and processes associated with analysing, designing and implementing ICT-based systems. The following aims of the AQA GCSE ICT syllabus indicate the nature of the subject:

- "Choose, use and design information and communication systems to carry out a range of tasks and to solve problems, making effective use of appropriate principles and techniques."
- "Develop a broad and balanced experience of the range of information and communication systems and their applications and an understanding of their capabilities and limitations." (Assessment and Qualifications Alliance 2000)

These courses typically have a substantial coursework element (normally accounting for 60% of the assessment) that focuses on solving problems through the use of ICT. This emphasis on problem solving through the use of ICT is also central to informatics courses (Hubwieser and Friedrich 1997).

For the GCSE in ICT students are expected to document the analysis, design, implementation and evaluation of their systems. A brilliant implementation without evidence of other elements of the process is only eligible for a very small percentage of marks. The educational purpose is to generate understanding of the nature of development of ICT systems (the systems development life cycle) by ICT professionals and to learn the skills, processes and strategies involved for solving simple problems that model tasks the professionals perform. Hubwieser and Friedrich emphasise the importance of the teacher choosing suitable problems as close to the student's experience as possible and encouraging all students to learn and use appropriate concepts of representation and techniques for carrying out the stages of the systems development life cycle.

The requirements for courses for 11-13 year-olds are less clearly specified (Webb, in press) but some schemes of work do recognise the need for understanding of the systems development life cycle and the development of the above-mentioned concepts and techniques (ACITT 1998; QCA 2000). The ACITT scheme introduces the systems life cycle in the first module of their course at age 11. The QCA scheme is less clear about the need to underpin the curriculum with the systems approach, only providing an overview of the systems life cycle in their final module - normally tackled by 13 year-olds.

ICT courses in England then emphasise practical problem solving through the use of ICT and the development of the knowledge, skills and processes that are required for these activities. There remains a lack of clarity about how progression in understanding should develop for 11-13 year-olds.

3. THE LEARNING PROCESS

3.1 Problem solving

There is no generally accepted model of problem solving that characterises the process. The CRESST model defines the main components of problem solving and incorporates many issues identified in a review of problem solving, predominantly from cognitive science research (O'Neil 1999).

The definition of problem solving used for this model is:

"Problem solving is cognitive processing directed at achieving a goal when no solution method is obvious to the problem solver." (Mayer and Wittrock 1996)

This definition can be applied to a wide range of problems but to make it really useful it is necessary to interpret each component specifically in relation to ICT-based problem solving.

An advantage of the model, shown in Figure 1, is that it reveals the wide range of understanding and skills required to solve problems and hence points to the complexity of teaching ICT where a predominant focus is on solving real problems. By the time students tackle GCSE coursework they need to be competent in all aspects of the model as they apply to ICT-based problems.

3.2 Content understanding

Content understanding is an integral component of the CRESST model of problem solving. Numerous studies that compared people who are competent in solving problems in various domains with beginners who are less proficient, found that competent problem-solvers are able to apply relevant principles from their content knowledge to the analysis and solution of problems.

What content understanding is needed for ICT? When and how should it be developed? Content understanding as specified by Hubwieser and Friedrich and GCSE syllabuses (Assessment and Qualifications Alliance, 2000) can be classified into:

- concepts and techniques of representation of data, knowledge and processes;
- metaphors and capabilities of types of application software;
- structure and functions of ICT systems;
- social, legal and ethical issues.

I believe there are differences in the approaches that are appropriate for teaching and learning of these four types of content understanding. The first two types are particularly critical in the problem-solving activities in which students engage for their coursework. Both types of content understanding incorporate declarative and procedural knowledge. The latter can only be learned through practice in problem-solving situations according to the "ACT theory of learning and problem solving" (Anderson et al 1995). According to this theory, declarative knowledge (e.g. that dataflow diagrams consisting of flows, inputs, outputs and stores can be used to model the logical flow of data around a system) is relatively easily taught and learnt probably by instructionist methods. The procedural knowledge that enables students to construct dataflow diagrams from descriptions of systems can only be learnt through practice. This will be obvious to ICT teachers for whom a typical teaching sequence on this topic might be:

1. The teacher explains concepts of dataflow diagrams and models an everyday example as well as a standard data-processing example.

2. The students construct simple dataflow diagrams of everyday examples.

3. The students draw dataflow diagrams for modelling parts of their designs for their coursework.

Metaphors and capabilities of types of application software also include both declarative and procedural knowledge. However, for developing understanding of this content, as well as emphasizing the importance of practice, I suggest benefits may be gained by adopting minimalism (Carroll 1998).

Minimalism is described as an action and task-based approach to instruction that emphasises the importance of realistic activities and experiences and focuses on errors for effective learning. Whilst it is not always possible to perform real tasks as they may be too complex, minimalist design of teaching and learning tasks would include elements of real tasks and provide scaffolding to enable learners to achieve them. The nature of the tasks would contribute to learners' motivation. An emphasis on getting learners started quickly on practical activities does not mean theoretical concepts are ignored - rather they are developed at appropriate points during a practical task. This approach appears to match what many ICT teachers strive to achieve, although the latter may not have defined their design principles or framework.

Carroll describes the minimalist movement as more a matter of zeitgeist and refinement than of radical innovation. The minimalist approach derives predominantly from writing software guides for adults but its use has extended into many other areas. It is underpinned by learning theory including Piaget's theory of learners as active constructors of their knowledge and Bruner's concept of scaffolding. Although minimalism suggests an approach that is simple and straightforward, it must be stressed that while the aim is to make the learning process simple and achievable, considerable effort and expertise is required to design learning situations that adhere to minimalist principles.

Much needs to be done in adapting minimalism to ICT teaching. However the principles, while grounded in popular learning theory, provide a more practical and focused basis for developing pedagogy than the general calls for a constructivist approach to teaching ICT. Little research or development has looked at the relationship between paper-based support materials, electronic support materials and teacher support. This issue should be addressed when establishing pedagogy for ICT.

3.3 Problem solving strategies

Domain-specific problem solving strategies may develop through the teaching of content understanding of representational techniques as long as the teaching and learning situation provides practice in their use. The use of a representational technique, such as a systems flowchart to model a problem solution, is a domain-specific problem solving strategy.

Domain independent strategies - the use of analogy or identification of boundaries of problems - can only be understood through tackling problems and focusing on the strategies used. There is a key role for the teacher in modelling strategies and scaffolding students while they are developing strategies.

The process of transfer of problem solving strategies is not always successful. Mayer and Wittrock argue that students may fail to recognise similarities between known and unknown problems unless there are explicit explanations and examples of similarly worked out problems. They consider it important that the students are initially able to convert the "problem-as-presented" into "an internal representation of the problem". Their research indicates that the student must develop cognitive and metacognitive strategies to accomplish this transfer of understanding.

4. SELF-REGULATION

Boekaerts (1997) summarises evidence from a range of studies that show that students who have more metacognitive knowledge in relation to a subject-matter domain, demonstrate superior strategy use and are better problem solvers. She concludes that self-regulation can be taught but that pedagogical strategies, including modelling, scaffolding and coaching need further research. With regard to metacognition I suggest ICT teachers should build on successful approaches in other subject areas. In particular the approach of Cognitive Acceleration in Science Education (CASE) might be adapted to ICT education (Adey 1999).

CASE activities are founded in theories of cognitive development described by Piaget and by Vygotsky. They are based on the idea that the solution of problems, with carefully graded help (mostly through questioning) from a teacher or more able peer, leads not only to a solution of that problem but also to the general stimulation of the pupils' cognitive processing mechanism. A carefully structured series of such experiences will have a permanent and irreversible effect on the subject's cognitive development. This is borne out by significant improvements in achievement of CASE pupils compared with control groups not just in Science (where CASE-intervention took place) but also in English and Mathematics.

CASE activities are tightly-focussed problem-solving tasks designed to develop understanding of scientific concepts related to Piagetian formal operations that students find difficult to grasp. In order to design similar activities for ICT it will be necessary to identify specific cognitive processes that students find difficult in ICT-based problems. Some processes may be drawn from those identified by the CASE project such as the understanding and use of variables, values and relationships.

CASE activities emphasise the importance of reflection, and in particular of social exchange, in the development of thinking as well as the development of knowledge. Pupils are encouraged to talk with the teacher or each other about how they are tackling and solving the problem or what difficulties they are finding. They become more conscious of their own thinking processes. This promotes cognitive development.

Two other important ideas used in the CASE materials are concrete preparation and bridging. Concrete preparation is the first phase of the activity when key words are introduced and practised and the context of the problem situation is made familiar through demonstration and questioning. The final phase of the activity is called bridging. In this process links are built between the type of reasoning being developed during the thinking activity and the use of that reasoning in other contexts. This bridging enables transfer to occur.

5. CONCLUSIONS

Problem solving is a large and essential component of the ICT curriculum in schools. The CRESST model of problem solving reveals the range of content understanding, skills and techniques that students need to learn to become effective problem solvers. I have argued that two key elements of content understanding required for ICT-based problem solving are concepts and techniques of representation of data, knowledge and processes and metaphors and capabilities of types of application software.

These may be best addressed by pedagogies based on Anderson's theory and on Carroll's minimalism respectively. The development of problem solving strategies depends on practice and appropriate scaffolding as well as techniques to develop cognition. A potentially fruitful approach is to develop a methodology based on that of CASE that would involve identifying key cognitive skills that students find difficult when solving problems in ICT and devising focused tasks.

REFERENCES

ACITT (1998) *Informatics Key Stage 3*. ACITT , London.

Adey, P. S. (1999) Thinking science as a gateway to general thinking ability. In *Teaching and learning thinking skills*. J. H. M. Hamers, J. E. H. Van Luit and B. Csapo (eds.), Swets & Zeitlinger, Netherlands, pp. 63-80.

Adey, P. (2001) 160 years of science education: an uncertain link between theory and practice. *School Science Review*, March 2001, 82. pp41-48.

Anderson, J. R., Corbett, A. T., Koedinger, K., and Pelletier, R. (1995) Cognitive tutors: Lessons learned. *The Journal of Learning Sciences*, 4, pp.167-207.

[http://act.psy.cmu.edu/ACT/papers/Lessons_Learned-abs.html]

Assessment and Qualifications Alliance (2000) *General Certificate of Secondary Education - ICT*. AQA Guildford.

Boekaerts, M. (1997) Self-regulated learning: a new concept embraced by researchers, policy makers, educators, teachers and students. *Learning and Instruction*, Vol 7. No 2. Pp. 161-186.

Carroll, J. M., (ed.) (1998) *Minimalism Beyond the Nurnberg Funnel*. MIT Press, Cambridge, Mass.

Hubwieser, P. and Friedrich, S. (1997) Teaching informatics as a subject. In *Capacity Building for IT in Education in Developing Countries*, London, G. Marshall and M. Ruohohen (eds.), IFIP Working Conference, Harare, Zimbabwe. Chapman & Hall, London.

Mayer, R.E. and Wittrock, M.C. (1996) Problem-Solving Transfer. In *Handbook of Educational Psychology*, D. C. Berliner and R. C. Calfee (eds.), Simon & Schuster Macmillan, New York. pp. 47-62.

O'Neil, Jr, H.F. (1999) Perspectives on computer-based performance assessment of problem solvlng. *Computers in Human Behaviour*, 15, pp. 255-268.

QCA (2000) *Information and communication technology: A scheme of work for Key stage 3*. Qualifications and Curriculum Authority, London.

Webb, M. E. (in press) *Pedagogical reasoning: issues and solutions for the teaching and learning of ICT in secondary schools*.

Development of multimedia animations – a contribution of informatics teaching to media studies

Michael Weigend
Fernuniversität Hagen, Holzkamp-Gesamtschule Witten, Germany

michael.weigend@fernuni-hagen.de

Key words: Curriculum changes, Informatics, Media Education, Problem Solving

Abstract: The development of a multimedia visual model can be an informative process that requires and encourages a large number of abilities. This paper focuses on three dimensions of the modelling process, all connected with certain abilities. Firstly, an informative model is a simplified image of a real system. In this respect modellers are required to observe reality well, to recognise and describe structures. Secondly, the model can be regarded as a medium through which certain mental constructs and contents can be imparted to recipients (communicative aspect). Finally, a multimedia model (especially if it is to be attractive) is a complex, technical product whose implementation demands the use of methods of information technology, for example object-oriented modelling. Using the example of the object-oriented modelling of a waterworks with the help of Flash 5, the paper discusses under what conditions the development of animation in IT teaching can help modelling abilities.

1. MODEL FORMATION IN SCHOOL

The development of models to explain real phenomena is very important, particularly in the teaching of science. There are detailed concrete teaching suggestions (see e.g. Ossimitz 2001) which are often based on the following pattern. On the basis of observations of a realistic, dynamic system (e.g. a forest ecosystem) a colloquial word model is formulated. This is then developed into a mathematical model and a computer simulation. In simulation runs with different input data the behaviour of the program is compared with the behaviour of the real system. If differences are discovered the real system is studied again in more detail and the model is changed. The model forming cycle continues until the behaviour of the model conforms well with the observed behaviour of the real system (Bossel 1987).

This process is supported by software tools such as Dynasys that allow the input of a model, a scenario of starting values and the output of a system's behaviour in a simulation. One weakness of existing systems is the limited demand on the creativity and imagination of the pupils. The possibilities in the definition of a model are basically limited to the arrangement of given symbols. The output during the simulation consists mainly of time and phase diagrams.

In the leisure market many successful simulation programs follow a common principle. Players build an artificial world from components (in SimCity they build a town with streets, shops, hospitals and industrial areas) which operates according to certain prescribed rules. The (sensory) attraction is in the observation of the effects of changes (e.g. the increase in tax yield because the town has highly successful industrial areas and has been able to draw in investors).

Programs of this kind are attractive and are used in many American schools as a medium for teaching pupils about complex systems such as social systems or ecological systems (Diener et al 1998). One disadvantage is that the logical model on which the system is based is not revealed to the player and cannot be changed.

As an alternative to the two above-mentioned approaches, this paper proposes the design of a visual model as multimedia interactive software (animation) with the help of Flash 5. In two respects we go beyond the modelling of dynamic systems suggested by Bossel. Firstly, questions of the communication design are included. The target product - the model - is seen as a medium conceived for a certain group of recipients. Secondly, in the model formation process, aspects of the technical realisability are taken into consideration.

An animation is a relatively complex system which can only be developed using information technology methodology. While there are many relevant informatic modelling techniques available (Hubwieser 1999), here we only consider the development of a class structure within the framework of object-oriented modelling.

2. ANALYSIS OF A WINDOW FROM REALITY - EXAMPLE: THE WITTEN WATERWORKS

It is an advantage if the "window on reality" has the following features - observability, structural feasibility and sufficient size. As an example, consider the modelling of a waterworks which was undertaken by pupils in years 9 to 11 at Holzkamp comprehensive school in Witten as part of a project week. (Detailed documentation, including the Flash animations mentioned in this paper are at http://www.informatik.fernuni-hagen.de/schulinformatik/projekt43/).

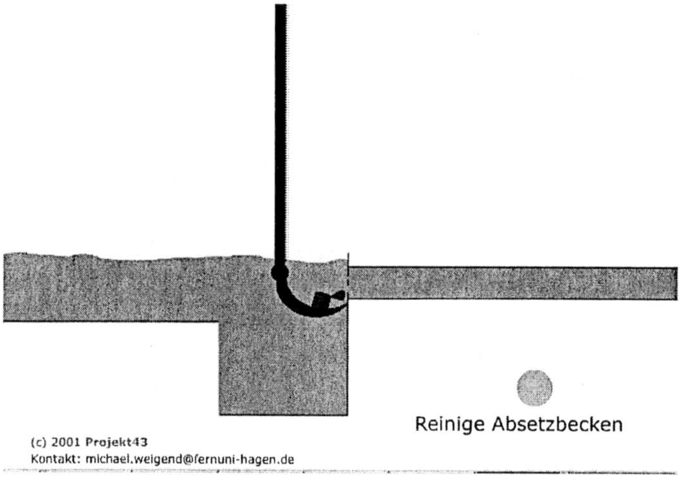

Figure 1. Flash animation screenshot of the fine screen system

The aim of the project was the creation of a multimedia animation with interactive, dynamic, visual and audio components - important components such as the fine screen which removes (strains) coarse particles from the river water and a settling pond that is occasionally cleaned by means of a mechanical gripper (Figure 1).

What makes a waterworks a suitable object for modelling?

- *Observability.* The window to be modelled is accessible to sensory perception and independent exploration. The waterworks can be visited. Its components can be photographed and measured. Original sounds of rushing water, pumps and compressed air can be recorded and incorporated into the animation. There are experts whom the pupils can question about details that are not immediately visible or evident (such as the depth of the rapid filter basin). It is only under these conditions

that model limits and reality can be discussed. The modelling project can also serve as motivation for pupils to investigate a realistic system (very relevant to their education) on their own.
- *Structural feasibility.* In order for the modelling task to be shared among teams it is advantageous if, at the top level of the hierarchy, the system which has to be described can easily be divided up into components (themselves subsystems) which can be modelled, as far as possible, independently. The structure of a waterworks is obvious and straightforward. The raw river water is prepared in several systems located one after another - rough cleansing with the screen, the rapid filter system, the slow filter basin, sterilisation with chlorine dioxide. Each team takes over a component and prepares its own demonstration model.
- *Size.* In order for pupils to experience the benefits of informatics methods of modelling, the system to be modelled must be sufficiently large and consist of many components. This criterion is certainly satisfied by a waterworks. There is little sense, for example, in using object-oriented analysis and design methods such as class, object and sequence diagrams if the task is only a matter of modelling a bouncing ball.

3. ASPECTS OF MODELLING

3.1 The problem of reduction

A model is a simplified image of reality. This means that in modelling we have to decide what aspects of reality we consider are essential and which ones we wish to ignore. The reasons for reduction are primarily:

- Limited resources (development time, competence of the development team, possibilities or capabilities of development software).
- Limited knowledge of reality.
- Limited perceptive faculties of the recipients of a model.

In a school project the first point is decisive. A project can easily fail if too much is taken on at the start. This means that at the beginning of the development process the pupils have to estimate how much time is required to realise certain features. They cannot do this, of course, when they are just beginners. The problem can, for example, be checked by an evolutionary or incremental development where a minimal system is first implemented and later expanded (Sommerville 1996).

A criterion for the selection of the elements to be presented in the model may be their significance for the whole system. The modelling of a filter process will hardly be a success without the presentation of running water! In this case the decision is based on the observation of reality.

From the aspect of media design it may be wise to consider including an unimportant aspect in the model simply because it can be visualised spectacularly. In water treatment the purification of the settling basin by a mechanical gripper is a marginal problem. On the other hand this process (in comparison to running water) looks interesting and makes the whole model worth viewing.

3.2 Modelling as a process of media design

At the end of the last section we mentioned that a model is also a medium for the communication of intellectual content. This means that in the design the addressee (or viewer) must be considered. Aesthetic preferences are important as are perceptive faculties and habits of thought - e.g. if the model is for pre-school age children it should not contain any written explanations. The addressee may also be the author of the model. In this case he/she invents models to make something clear to him/herself. The communication aspect is obvious in models we know of from everyday life:

- Images and models of buildings built by architects for decision-makers.
- Model-based scientific demonstrations at school.
- Didacticised explanatory models e.g. Dalton's model that explains aggregate states.

Under the perspective of communication design the following points are relevant and are discussed with respect to the waterworks example:

- Graphic design. This includes the selection of the colours and the transparency. Important parts of the model (the floating can and the gripper) should stand out optically.
- Interaction. The functioning of the fine screen can be best illustrated using an automatic simulation. For the observer, however, it is more interesting and instructive if he or she can press a button and set the machinery working.
- Selection of symbols and pictures of concrete objects. Here an empty can was chosen as an example of an object that is removed from the fine screen after being in the raw water. A floating branch or leaves would actually have been more realistic but a can is a symbol of how man pollutes water and the addressee is more shocked.

- Design of the user interface.
- Illustration of movement. One particular problem was the visualisation of flowing water. Pupils solved this by making little bubbles float through the canal.
- Reduced illustration. An economic approach to development time is essential and results in a design problem. How can a particular motion process be illustrated as simply as possible? The mechanics of cleaning the basin in the example just consists of two moving parts.
- Exemplary processes. For the representation of a whole class of events in the real world, a single typical sequence must be selected. At a real waterworks very different objects float against the fine screen in different ways. In the animation the movement of a can was visualised to represent this.

3.3 Modelling as an experiment

What has been said so far throws light on the complexity and size of the problem solving processes associated with modelling. Problem solving is described by cognitive psychology as cyclical (Anderson 1996). In the search for a way from the initial state to the target the person attempting to solve the problem tries out various operators and checks whether or not they bring him closer to the target. Designers and software developers can work through parts of these cycles in their minds. Beginners do not have this experience or ability and thus need a lot of opportunities to experiment. An animated film sequence is first prepared as an experiment. In test runs the optical effect is assessed. Weak points are identified and improvements are made until the result is satisfactory.

In the waterworks project one pupil spent a whole day visualising the undulatory motion of water in a basin. He developed more than a dozen variants until he found a solution he considered acceptable. Experimenting like this develops the ability to solve problems. In a modelling project in which several persons co-operate two conditions must be met so that there is room for independent, individual experimentation.

First of all the development environment must allow pupils to prepare the model bit by bit and they must be able to test each part individually. Moreover there should be support for the rapid design of dynamic, graphic objects with no syntax errors. Macromedia Flash provides this. The definition of the course of movement of a graphic object through a "tween" usually takes seconds. Pupils can therefore go through hundreds of development cycles during a modelling project.

In the implementation of graphic software with Java using class libraries such as AWT or Swing, experimentation is more difficult and time-consuming. Teaching suggestions for modelling which plan an implementation in Java often assume that experimentation has finished before the first line of the program is written. Here the actual model formation is done almost exclusively with pencil and paper during the phases of analysis and design.

The second condition for individual experimentation concerns the system design. The software developed must meet the quality criterion of "changeability" to a high degree. This means that a change in one part must not have any effects on the whole system.

3.4 Modelling as the development of a class structure

In view of the short time available to a school project, a multimedia model is a complex, technical product that can only be managed by dividing up tasks in a team. The basis for the structuring of the development work is system design. If the object-oriented modelling paradigm is followed the program structure is represented through a system of classes (Balzert 1999). With visual software each class usually describes the features of objects visible on the monitor.

Using the UML (Unified Modelling Language) the students construct class diagrams for different parts of the Waterworks model. The classes are represented through boxes with the class names and connecting lines are used to show the relationships between the classes.

For the design of animations aggregations are very important. This concerns the relationship between an entirety (an aggregate) and its parts. For example, in one class the cleaning mechanism is defined as "Excavator". This aggregate consists of the classes "Rod" and "Grabber" in which the dynamic behaviour of the two components of the excavator are described.

In the class "Grabber", however, an object of the class "PictureGrabber" is used. This represents a static image used in the higher class "Grabber" for the definition of a film sequence. If the picture of the grabber is to be replaced later by a more realistic image, it is sufficient to just change the class "PictureGrabber". All other classes are completely unaffected.

The class "Bubble" which describes a moving (seemingly drifting in the water) air bubble only has to be described once. In the development of the model many objects (instances) can be used by a class to represent flowing water.

The division of tasks - i.e. the allocation of classes to persons who will implement them - can be done according to different criteria. One of them is the specialist knowledge of the respective pupils. Anyone with a special talent for drawing should be responsible for the design of the more difficult static images. It also makes sense if one person takes over a whole sub-tree of the class aggregation. In this case any later changes to the design within this area have no effect on the work of other team members.

The examples illustrate the main advantages of a good design:

- Time is saved by the multiple use of one class.
- Advance structuring of the work process.
- Achievement of changeability.

Finding a suitable class structure is a complex decision process. With visual programs it is made easier by the fact that the analysis of the reality to be illustrated already provides assistance with the structured division of the whole system. Nevertheless a class diagram of the model does not always have to be a faithful structure of the original in reality. This means the components of the software may contradict the real facts.

Figure 2. Erwin the duck screenshot

The following example comes from another project. In the game "Erwin the duck" a flying duck has to be guided past poisonous clouds of smoke coming from factory chimneys (Figure 2). If the duck touches a cloud it becomes unconscious and plunges to the ground.

One problem the pupils faced was the modelling of the factory chimney. They had many attempts at a class diagram and their efforts were guided by the structure of the real system. The factory is an aggregate from the building, the chimney and a dynamic cloud that keeps coming out of the chimney and is carried away by the wind.

The problem with the implementation using Flash was that the chimney was drawn in perspective and the upper opening was represented as flat and oval. It always looked as if the cloud came out from behind the chimney - not from the opening. The problem could only be resolved by breaking away from the structure of the real system and representing the chimney using two separate objects - a front and a rear. The cloud could now be positioned between them.

In this case we had to take into account the fact that Flash only supports two-dimensional graphics. In an implementation with VRML (Virtual Reality Modelling Language) the chimney-cloud problem would not have occurred. The beginning of the cloud could have been put in the interior of a three-dimensional chimney.

Note that the development of a class diagram in the framework of an object-oriented design is a creative process. It includes the facts gained from an analysis of reality and the knowledge of the possibilities of implementation. For this knowledge to be gained it is necessary for development cycles to be gone through as frequently as possible. In particular this means that modelling must not get stuck in the design phase - design errors are often not obvious until after implementation.

4. CONCLUSION

What Karel the Robot and LOGO turtle graphics were and are for learning algorithms, tools for preparing interactive animations could be for modelling complex systems. They are platforms that support experimentation and the acquisition of knowledge through experience. In modern society animations are an important part of media culture. Their construction and reflection should therefore be part of media teaching in schools.

REFERENCES

Anderson, J. (1996) *Kognitive Psychologie.* Oxford , Heidelberg, Berlin.

Balzert, H. (1999) *Lehrbuch der Objektmodellierung. Analyse und Entwurf.* Heidelberg Berlin.

Bossel, (1987) *Systemdynamik.* Braunschweig.

Diener, U., Dönhoff, H-U., Rieks, K-E. and Weigend, M. (1998) *Neue Medien im Unterricht - Vorbild USA?* (Verlag Bertelsmann Stiftung), Gütersloh.

Hubwieser, P. (1999) Modellierung in der Schulinformatik. In: *LOG IN* 1 1999, S.

Ossimitz, G. (2001) *Didaktik der Systemdynamik, systemisches Denken.* Universität Klagenfurt.

[http://www.uni-klu.ac.at/users/gossimit/sdyn/sdyn.htm.]
Sommerville, I. (1997) *Software Engineering*. Addison-Wesley.

ICT: an aid to inclusion? Reflections on the potential of ICT for the changing role of the special school

Chris Abbott and John Galloway

Department of Education and Professional Studies, King's College London, Franklin-Wilkins Building, London, SE1 9NN, United Kingdom. chris.abbott@kcl.ac.uk

SEN Centre, 85 Harford Street, London, E1 4PY, United Kingdom. jgalloway@towerhamlets-clc.org.uk

Key words: Agent of Change, Classroom, Disability, Equity, Special Needs

Abstract: Inclusion has been an aim for many national education systems but progress has, in some countries, been slow or halting. Information and Communication Technology (ICT) has been a key factor in the greater inclusion of individuals with special educational needs, either by enabling them to communicate more readily and widely, or by assisting them to overcome their learning difficulties. The paper examines the extent to which ICT can contribute to social change by redefining the role of the special school. A case study of one special school, given a range of support and resources, attempts to discover the potential of ICT in this area. If a school is provided with appropriate hardware and software, training in ICT for all teachers, and the support of a specialised advisory teacher, can that school develop valuable resources for the wider education community? How should these resources be disseminated? The progress of the case study is described through the use of semi-structured interviews, diaries and commentary, and a number of emerging issues are identified.

1. INTRODUCTION

This paper describes a small-scale intervention case study which took place over nine months in an inner-city area of the UK and relates the findings to wider issues of inclusion, ICT and social change.

Over the last twenty years most education systems in Europe, the USA and Australasia have begun to implement some form of inclusive education. In a number of cases, this has meant providing access to education where none was previously available; in others the focus has been on extending access to education to groups not previously offered this opportunity.

Current UK policy is firmly aimed at inclusion, although precisely what this means can be a matter of interpretation. Inclusion has often been in conflict with other policies, especially those regarding school standards and achievement (Abbott 2001). Despite this, government policy in the UK has been only one driver towards inclusion, with individuals and organisations deploring the methodology of separate provision - even the 'pathologising' of children (Billington 2000).

With the implementation of the UK National Curriculum in 1988 it was expected that some students with severe learning difficulties would be dis-applied from the curriculum. In the event, this rarely took place and the National Curriculum, and access to it, became an entitlement issue. At the same time, special school closures continued. One *mixed needs* school often replaced three or four alternative and specialised schools or units. The *mixed needs* model is clearly less expensive. Despite the closures being particularly marked in areas with historical reasons for greater special education provision (such as London), the percentage of UK children in separate provision has changed only slowly. In 1978, 2% of the school population was in special schools; by 2000 this had dropped to 1.3%.

Where special schools have survived, they have often changed their role and adopted new and innovative practices. Our case study examines one such school, catering for students who attend full time from the ages of 2 to 11 years and who have severe or profound and multiple learning difficulties. The school, which we will call Eastville, is an award-winning, purpose-built school - barely five years old. It currently has 71 pupils and a staffing ratio of almost one to one. Teaching staff are supplemented by teaching assistants and supported by other services including speech and language therapists and physiotherapists.

Eastville School is in an inner city area which has pockets of severe deprivation punctuated by expensive housing for the professionals who work in the expanding service industries. Over 50% of Northstone LEA's school population are from ethnic minorities.

2. THE CONTEXT – ICT AND SPECIAL EDUCATIONAL NEEDS

It has long been recognised (McKeown 2000, Abbott 2000a, 2000b, 2002) that ICT can play a major role in assisting students with special education needs (SEN). It can enable them to communicate with others, become fully literate and learn more effectively. However, the ICT orders in the UK National Curriculum may not be accessible to all students with SEN, especially those who are likely to spend most of their school career working towards the lowest levels of the curriculum.

For this reason, guidance was produced (QCA 2001) to provide exemplar and other material in ICT and all other subjects with this target group in mind. The guidance included performance descriptions known as the *P scales* or *P levels* offered for performance across and within subjects. They describe performance before Level 2 of the National Curriculum in eight stages, from P1 to P8. For example, P4 for ICT reads:

"*Pupils intentionally make selections to communicate meanings. They know that certain actions produce predictable results, and that ICT can be used to control objects, events and aspects of the environment.*" (QCA 2001: 19)

It has been suggested (Colwill and Peacey 2001) that the National Curriculum is not a curriculum at all and that the curriculum delivered by the school is what matters. It is good to see an emphasis on teachers and schools "*determining and constructing their own curriculum*" even if that might not be how some of those teachers see it.

"*Throughout the ten years of the National Curriculum many teachers working with pupils with learning difficulties have adapted the National Curriculum for use in their situations. There has been no commonly agreed approach to the adaptations and no articulation of a curriculum model either to assist schools in curriculum development or to provide a blueprint against which schools' performance can be judged.*" (Colwill and Peacey 2001: 122)

The tension would seem to be between the "*commonly agreed approach*" and the teachers who are busily determining and "*constructing their own curriculum*". Initial reactions from Eastville School were not particularly positive.

"*Well, we've started looking at this, the Advisory Teacher and me, but we've also gone back to the actual National Curriculum. We've actually been using that. And we actually in the end found that more of a help than this.*" ICT Co-ordinator Month 2

3. WHAT HELPS? DESCRIPTION OF THE CASE STUDY

Eastville, like all schools in the UK, has received considerable extra ICT resources since 2000 through the Department for Education and the Environment (DfEE) (now Department for Education and Skills (DfES)) Standards Fund. All its computers have been purchased in the last five years, since the school opened. It has thirteen computers for curriculum-oriented work, a well-equipped sensory room, and a new music room with a skilled part-time technician. So far, funds have gone into increasing the available technology: in the next financial year funds will have to begin to address the problem of replacing outdated equipment.

The school has also been provided with a generic suite of software (Infant Video Toolkit – 2Simple Software), originally designed for younger children, which could offer an appropriate resource for older students with severe learning difficulties. Experience in the classroom and training they have received has made the teachers clear about what makes software appropriate for their students.

"something that's visually not cluttered even if the actual operation of the program is relatively simple." Year 5 Teacher Month 2

Eastville School, like all UK schools, received staff in-service ICT training supported by the New Opportunities Fund (using proceeds from the UK National Lottery). This training was purchased from a prescribed list of specialist SEN providers and staff considered the training was of value.

"*Lots of good ideas... especially as I was currently new to IT and special needs. But I think I need more. Well I know I need more...*" Year 4 teacher Month 2

"*What I found very useful was the huge wad of notes and information. I haven't read it all but it's quite useful as a resource to go to...*" ICT Co-ordinator Month 2

The school also had access to the equivalent of nine days of support from a designated ICT and SEN advisory teacher. This was given as extra training, classroom support and by working with individual teachers. A particular problem for a school offered this kind of support is deciding how to make best use of it.

"*I think spread out regular... But say you're working on a particular program* (having an Advisory Teacher) *gives you... time to work with the children, especially this type of child that may need a lot of practice.*" Year 5 teacher Month 2

Eastville School therefore has the four components needed to enable effective teaching of and with ICT:

- In-service training for teachers.
- Appropriate software.
- Advisory teacher support.
- Guidance documents and resources for curriculum delivery.

The case study examined the effect of each of these support mechanisms on the process of developing an ICT Scheme of Work (SoW) in the school for Key Stage 2 (children aged from 8 to 11 years). Semi-structured interviews with two classroom teachers and the school's ICT Co-ordinator took place at the beginning and end of a nine-month period during one school year. The Advisory Teacher and the Co-ordinator kept diaries of progress and, during early meetings in Month 1, they developed the initial SoW. They quickly realised that a suitable style to adopt would make use of the phrase "*All/Most/Some pupils will be able to*" - based on terminology in curriculum documents (QCA 2001).

There were also discussions about the use of e-mail, the pupils' understanding of e-mail, the needs of 'switch' users and the usefulness of existing resources. During Month 2 the focus was on the use of existing materials alongside the school's Development Plan and other internal documents. It was decided that ICT would be taught as a discrete subject at Key Stage 1 (KS1), although this age-group had not formed part of the original plans. Much discussion around mechanisms for sharing useful Web resources resulted in a mix of training sessions for staff and guided use of the staffroom computer and its book-marked sites. A group of staff evaluated Infant Video Toolkit and concluded that many elements of the package would meet pupil or staff needs.

Next, activities leading to Logo became a focus and a variety of radio-controlled toys were purchased to develop concepts of control, direction and speed. During Month 3, school responses to symbol e-mail were discussed with the Advisory Teacher and further sections of the SoW were completed. Aspects of the use of one computer in a classroom were discussed and different formats of lesson plans were considered.

By Month 4, the ICT Co-ordinator had developed lesson plans for all three areas of the SoW and all four years of KS2 - a much greater achievement than had been expected. The Co-ordinator also produced an ICT plan highlighting activity for the remainder of the academic year. She planned twilight training sessions to introduce the lesson plans and asked all teachers to try out some of the lessons by Month 7. At this point KS1 was added to the SoW and thus a comprehensive programme covering the whole of primary education from 5 to 11 years was developed. By Month 8, the school had begun to look at existing ICT assessment schemes aimed at people with learning difficulties. The SoW was now tied closely to the P level guidance. It was agreed that it would be introduced from the beginning of the following academic year (Month 11) following ratification by the School Governors.

4. ISSUES ARISING FROM THE CASE STUDY

A number of issues have arisen. Many have been identified before (Abbott and Cribb 2001) and the case study provides further evidence.

4.1 Individual schools versus national initiatives

One issue relates to the effects of individual school initiatives versus what might be termed national diktat. When many older teachers began their careers they had almost total control over the curriculum taught in their classrooms and developed materials accordingly. An associated issue relates to resources developed for teachers rather than by them, and the extent to which these resources can be used effectively by the target group. It is striking to note the increased level of confidence shown by the staff at the end of the case study period. The ICT Co-ordinator had recognised that printed materials can only be useful as a starting point.

4.2 Isolated position of special schools

Special school isolation has been identified as part of the emerging research agenda around special schools (Abbott & Cribb 2001). At Eastville isolation was less marked than might have been the case without the presence of the Advisory Teacher and his ability to help staff locate themselves within their Local Education Authority and the wider SEN community. ICT and the World Wide Web can help schools avoid this isolation (Abbott 1998, Galloway 2002). Indeed, the ICT Co-ordinator, who is also the school's Early Years Co-ordinator, began working one day a week at a local primary school. This enabled a changed working relationship between school and LEA advisor, since both individuals now work alongside each other to support this mainstream school.

"*(Eastville) as a school in principle is very happy to share its schemes of work with anyone and the idea is that at some point it'll be on the website and people will be able to download them.*" ICT Co-ordinator Month 9

The placement was a recognition of the teacher's experience in the area of Early Years and ICT, and is another indicator of the ICT-related changes in working practice that are blurring the boundaries between mainstream schools. However, at the local level the lack of a special school ICT Co-ordinator network, common in the mainstream school system, continues to cause concern.

4.3 Role of resources in curriculum change

The usefulness of documentation such as the P-scales is clearly a key issue. They represent the newest resource available to the school and have been issued for all subjects. Perhaps the most significant positive outcome recorded by the participants was that they were able to decode the P-scales sufficiently to understand them fully and make appropriate decisions about their usefulness. Staff at Eastville and other special schools are also aware that, in future, they will be asked to set targets for all their students in some subjects. Our case study examined the extent to which the varying support considered might assist in this process.

4.4 From long-term to short-term placements

A further development in the changing role of the special school was the rapid development of short-term placements at Eastville. Towards the end of the project almost 40% of the school roll consisted of short term placement pupils in the nursery, most of whom were expected to progress to mainstream schools after a relatively short time in special education. There are now as many Early Years pupils at the school as there are in Key Stage 2. The expectation was that the needs of these very young children would be addressed and that Eastville would prepare them for a mainstream placement.

5. CONCLUSION

This paper has shown how ICT is involved in the process by which special schools could move more quickly to a wider resourcing role than most currently play. If one special school is able to use these resources and support mechanisms to develop an ICT scheme of work, could it support other schools and share what it has created? Must every school go through the same process, each in their separate and exclusive settings? Can the process used to develop a scheme of work in one curriculum area act as a model for other areas?

It is our assertion that ICT has a key role to play as an agent of change and to enable special schools to take on their role as a resource for the mainstream. This assertion is not based on any simplistic technological determinism, but through a combination of the status of ICT as a *new* subject, its power to offer crucial support for those with SEN and, not least, its intrinsic ability to attract funding from governments and education authorities.

REFERENCES

Abbott, C. (1998) *Making the Web Special*, King's College/Institute of Education, London.

Abbott, C. (2000a) *ICT: Changing Education*, RoutledgeFalmer, London.

Abbott, C. (Ed.) (2000b) *Symbols Now*, Widgit, Leamington Spa.

Abbott, C. (2001) Special educational needs: Becoming more inclusive, In *Becoming a Teacher: Issues in Secondary Teaching* (2nd ed.), J. Dillon, and M. Maguire, (eds.), pp. 192-202, Open University Press, Buckingham.

Abbott, C., and Cribb, A. (2001) *Special Schools, Inclusion and the World Wide Web - the emerging research agenda*, British Journal of Educational Technology, 32 (3), pp. 331-342.

Abbott, C. (2002) *SEN and the Internet: Issues in Inclusive Education*, RoutledgeFalmer, London.

Billington, T. (2000) *Separating, Losing and Excluding Children: Narratives of Difference*, RoutledgeFalmer, London.

Colwill, I., and Peacey, N. (2001), *Planning, teaching and assessing the curriculum for pupils with learning difficulties: curriculum guidelines to support the revised National Curriculum*, British Journal of Special Education, 28 (3), pp. 120-122.

Galloway, J. (2002) Developing an LEA website to support the inclusion process. In *SEN and the Internet: Issues in Inclusive Education*, C. Abbott (ed.), RoutledgeFalmer, London.

McKeown, S. (2000) *Unlocking Potential: How ICT Can Support Children with Special Educational Needs*, The Questions Publishing Company Ltd. Birmingham.

QCA (2001) *Planning, teaching and assessing the curriculum for pupils with learning difficulties: Information and communication technology*, Qualifications and Curriculum Authority, London.

Various modelling aspects of tutoring systems for people with auditory disabilities

Nelson Baloian and Wolfram Luther
Department of Computer Science, Universidad de Chile, Blanco Encalada 2120, Santiago, Chile, nbaloian@dcc.uchile.cl

Institut für Informatik und Interaktive Systeme, Gerhard-Mercator-Universität Duisburg, Lotharstraße 65, D-47048 Duisburg, Germany, luther@informatik.uni-duisburg.de

Key words: Disability, Learning Environments, Learning Models, Recommendations, Tutor

Abstract: Following a survey of the existing tutoring systems for hearing impaired people the paper describes common aspects of modelling the real world when creating educational software. This involves testing and evaluating cognitive tasks with people who dispose of reduced auditory cues. To validate our concepts we used the multimedia learning system known as Whisper. We developed Whisper in 1997 and it has been evaluated extensively in special schools for the hearing impaired. Members of all the leading German rehabilitation centres have assisted in this evaluation process. Whisper implements an interactive educational system to help hearing-impaired people recognise speech errors.

1. INTRODUCTION

A number of systems have been developed with the intention of being used by people with disabilities. Systems created for hearing-impaired persons are oriented to train people by developing the necessary skills to overcome their disabilities.

We recognize in these systems similarities with the process of modelling the real world in a way with which the impaired can explore and interact.

Mostly, this process consists of two steps: the first step involves modelling of the real world within the computer and the second projects this computer model as a multimedia type with which the impaired can interact.

In most systems the modelling process ends at this point. Systems supporting learning or skill acquisitions need a feedback with a meaningful response. That can only be provided if the system has an embedded student's model constructed upon the learner's answers to certain questions or even the requirements proposed by the system. In this article we address both issues, the modelling of the real world and of the learner.

We started developing a multimedia learning system in 1997 called Whisper (in both German and English). This system enhances the proficiency in speaking for postlingually deaf and in hearing for Cochlear implant users (Wans 1998; Wans 1999) Whisper offers a workplace for a hearing-impaired person to recognize speech errors. Words, sentences or small stories are taken from everyday life. The learner's best verbal productions spoken in the presence of a teacher or a hearing person are recorded and stored in a repository. The learner explores a typical situation presented by appropriate cartoons subtitled with names which are spelt using a phonetic alphabet. Then the learner has to repeat all names or short sentences. The utterances are recorded under similar conditions and an acoustic representation of the spoken words is compared with the patterns stored in the database. By means of intelligent pattern recognition algorithms the difference between the learner's best production and the current expression is evaluated against five important speech parameters. An appropriate graphic and textual output for the learner is generated. This visual feedback replaces the usual auditory outcome and enables the impaired person to improve their speech capabilities.

The use of Virtual Environments (VE) as an interactive technology to allow learners to explore, interpret and represent real and symbolic objects and actions is a promising tool for people with sensory disabilities. The virtual environment is a navigable world built by using appropriate modelling language like VRML 2.0, dynamic scene objects, and acting characters. Scenic objects are defined by graphic and acoustic attributes and the actions of each character are based on deterministic and non-deterministic plans described within an interactive hyperstory. According to Lumbreras and Sanchez (1999) a hyperstory occurs in a virtual environment as a hypermedia version of a literary story. A new version of Whisper will incorporate these ideas and allow the creation and acting out of short stories using special characters like cartoons.

This paper begins with a survey of existing systems for hearing-impaired people. Next we develop common modelling aspects leading to a model for Whisper and similar systems. This is followed by a discussion of all the important steps of the modelling pipeline - the reduction of the original model to only graphic objects, the reconstruction of the internal computer model by the learner, the integration of an error diagnosis together with a user-adapted output, and the evaluation process.

2. RELATED SYSTEMS FOR HEARING-IMPAIRED PEOPLE

Our main interest group of hearing-impaired are the postlingually deaf who are familiar with the terms of the language because they have lost their sense of hearing after the acquisition of language and speech. This group of hearing-impaired has to live with severe communicative restrictions. The missing auditory comment often causes some deterioration in the control of speech production impacting both the intelligibility and social acceptability of the deafened speaker's voice. For deaf people the sign and gesture language was developed to allow nonverbal communication. The language provides all finger spelled letters used for proper names and rare words as well as thousands of word signs formed by hand and arm motions. To provide emphasis and to structure sentences, head motions and facial expressions are used. It is actually possible to create a dialogue based on hand gestures such as the dialogue between deaf human and deaf virtual human. A real time facial animation engine conceived to run on a common PC with a graphic accelerator card is described in Breton et al (2001).

Mehida (Alonso et al 1995) is an intelligent multimedia system for deaf or hearing-impaired children designed to assist them in acquiring and developing communication skills. It covers the following types of communication - finger spelling (representing the letters of the alphabet using the fingers), gestures or sign languages, lip reading (understanding spoken language through observing lip motion), and voice recognition.

The IBM SpeechViewer III (IBM 2000) is a powerful speech and language tool that transforms spoken words and sounds into imaginative graphics. SpeechViewer III is intended for people who have speech, language or hearing disabilities. It offers three main types of exercises. Awareness exercises are simple games concentrating on basic speech parameters and can be used by very young children. The skill building exercises ask the learner to work towards achieving a goal set by the expert and provide feedback on pitch, voicing, and vowel articulation. Patterning exercises offer a graphic representation of the learner's and the expert's voices for comparison concerning pitch and loudness.

ISAEUS (1997) is a speech training system for deaf and hearing-impaired people on a multilingual basis which has been developed through an EU-project involving groups in France, Spain and Germany.

The Visual Talker (Visual Talker 1999) and a recently developed product from the Liberated Learning Project provide real-time speech-to-text visual translator systems for classroom environments. The aim of this research is the development of a real-time speech-to-text system to be used by teachers, their students with hearing loss and/or speech impairments, and all other students in classroom settings to provide a 'full participation' communication system.

In summary, we can identify two types of enhanced systems for people with sensory disabilities: systems that try to complement common interfaces with other communication facilities and systems that support direct interactions with internal models or adequate representations of virtual models. Both try to transform the audible output into a textual or graphic format through which the user should navigate or communicate. We will now show how the modelling process should reflect the user's impairments.

3. MODELLING PIPELINE FOR DEVELOPING EDUCATIONAL SOFTWARE

The modelling process starts with the definition of the cognitive skills the learner has to acquire; then it considers the creation of a virtual environment composed by a navigable world and built by using an adequate modelling language, dynamic scene objects, and acting characters. Scenic objects are characterized by graphic and acoustic attributes and the actions of each character are based on deterministic and non-deterministic plans as in an interactive hyperstory. The learner explores the virtual world by interacting with appropriate interfaces and immediately obtains feedback.

The learner's actions, such as sound reproductions or utterances, are collected, evaluated, and classified, based on a student's modelling and diagnostic subsystems

In Whisper the modelling process is as follows. First, a virtual model **B** (a graphic, textual and phonetic model) is derived from the real or ficticious world scenario **A** (of objects, phonemes, words and stories) by means of abstraction and reduction - without taking into consideration the limitations of the potential users. The modelled objects are nouns or short animated stories which are named using phonemes and words. Then model **B** is projected to an adequate model **C** (a visual, frequency colour model) which can be explored by people with sensory disabilities. Obviously certain information channels cannot be used. Appropriate editors support the modelling process and important model parameters must be identified at this stage. By interacting with the system, the learner makes an internal reconstruction of the model **C** called **D**. In Whisper, the learner explores the visual model of a short story and reconstructs the visual wave form representation by speaking words or sentences. Additionally, the learner may build an external representation of **D** which has to be evaluated. This can be done by using an appropriate multidimensional error measure $\mathbf{m_I}(\mathbf{P},\mathbf{X},\mathbf{Y})$ depending on the objects and their attributes, the parameters **I**, a learner **P**, the computer model **X**, the reconstructed model **Y**. Finally, the degree of similarity is derived from the error measure and the result is displayed in a learner-adapted output and used for updating the student model. The actual representation is checked by a human tutor or an intelligent computer module to look for any correspondence with the original model.

The error measurement (represented by the function $\mathbf{m_I}(\mathbf{P},\mathbf{X},\mathbf{Y})$) should reflect this difference. The index **I** stands for different properties of objects being a part of **X** and **Y**, **P** denoting the learner, **X** the internal model **C**, and **Y** the reconstructed model **C** consisting of the words or sentences. Then the 'distance' between the same object in models **C** and **D** will be calculated using an appropriate vector norm. Thus we can derive a candidate for an error measure $\mathbf{m}_{\text{Distance}}$.

3.1 Construction and evaluation of the internal model

Now to the issue of reconstructing the internal model and of finding out ways to measure how close to the computer model is the user's mental model. In learning oriented systems this task's goal is to give meaningful feedback to the user's performance. While in the presence of a human tutor, the reconstruction process is monitored and supervised. Practicing without assistance gives more independence and self-confidence, but requires some integration of student models together with diagnostic subsystems.

Studies on the speech production of the postlingually deaf have shown that the loss of auditory feedback affects the speech proficiency especially in the global parameters of articulation, accentuation (stress), intonation, pitch, rate, and volume. The most striking speech error involves the volume control. Because of the missing auditory feedback the postlingually deaf either speaks too softly or tends to shout.

The loss of the sense of hearing has also an important influence on articulation. In most cases, the sibilants are mis-produced. Another common type of error is slurring, which is elongation of segments with increased co-articulation giving the loss of the segment definition. Frequently, the prefix or suffix of a word is omitted, segments are added, omitted or substituted.

These errors tend to occur in clusters as if some aspect of the control of speech had temporarily become maladjusted. On the other hand, the articulation is sometimes exaggerated. The nature of speech determined by stress and intonation is often affected. Pitch is then monotonous, stresses are excessive or misplaced, or patterns of intonation are inappropriate. The intonation can be very individual - it may correspond neither to the sense nor the purpose of the statement and makes the statement hard to understand.

After each utterance the system has to find out the differences between the statement of the learner and the reference model in the database, analyse the reasons for the modifications, and provide information on how the learner can improve speaking. There are two important methods for analysing speech parameters and to detect errors - signal analysis or algorithms for speech recognition.

In the first case, the differences concerning the parameters of volume, articulation, stress, and intonation must be visualized. From these visual patterns important parameters are detected by means of picture processing analysis of, and in connection with, the stored reference patterns. An intelligent component of the system interprets these results, displays a visualization of the error function $\mathbf{m}_l(\mathbf{X},\mathbf{C},\mathbf{D})$, for each parameter **l**, and transforms them into instructions for the learner.

We sample the recorded speech signal at a rate of 22.05 kHz and 16 Bit encoding. Then we execute 40 fast Fourier transforms per second to obtain a frequency representation. For display we use an RGB colour map with three overlapping triangular characteristics for red, green, and blue in the intervals [0, 1000], [250, 4000], [1000, 10000] which can be freely modified. The resulting frequency-colour representation is smoothed and simplified by an appropriate cluster and threshold algorithm. Whereas the wave amplitude image provides in the xy-plane any feedback concerning volume and rate, omitted syllables are characterized by a lack of coloured areas, intonation and articulation by the parameters hue and saturation as well as the shape of the curves. Details are described in Hobohm (1993) and Gräfe (1998).

In recent years it has been shown as reported by Willett (2000) that the combination of neural networks with Hidden Markov Models (HMM) is a powerful approach to speech recognition. These so-called hybrid speech recognizers use multi-layer perceptron or recurrent neural nets to estimate the a posteriori probabilities for the phones in each feature. There are different new approaches to build three-state-phone HMMs in a hybrid recognizer, one state represents the beginning of the phone, one the static part of the phone, and the third state is used to model the transition into the following phone. When using context-dependent models, even small systems with a small vocabulary lead to an acceptable number of triphones in the output layer of the neural net. Additionally, this approach permits the combination of discriminative training techniques with Maximum Likelihood methods for the HMM parameters using a phoneme-labelled speech database. However, in our case for the training procedure on the learner's best performances, phonetic transcriptions are necessary. The system provides a morpheme-to-morpheme transposition of an oral language into the visual channel. In this process all structures of the special oral language are preserved. Whereas the system adapts to the speaker (taking in consideration the collected training data), it does not have the intention of reducing the error recognition rate. On the contrary, the system is intended to detect errors concerning deformed or omitted phones or erroneous articulation. The system searches most probable phone sequences and the phones are displayed within a separate window.

The combination of a well-known and efficient speech recognition architecture together with new techniques of speech error detection seems to be a promising research field in the future and will be used in the context of speech error detection in a new version of Whisper.

3.2 User adapted output

The natural way to communicate and to comment on the outcome of a session would be to introduce a human expert. However, permanent assistance is expensive and not always available.

Currently our technical frequency-colour representation of speech is not quite adequate for young users. It seems to be more motivating to use special icons or comics. We have adapted forms and colours following observed error parameters. For example we might create visualisations of a learner's best production of the word aubergine a current production of the French word 'aubergine' with corresponding intonation error and add to the visual feedback a simplified representation of an aubergine as a cartoon which cleverly includes the most important features of the technical output. Then we give the following guidelines:

- Simplify the envelope of the speech pattern.
- Use the time and volume axes to indicate these errors by a simple transformation of the cartoon.
- Articulation errors can be represented by deformation of the contour.
- Omission errors are modelled by omission of control points in the B-spline representation of the graphic object.

Our task is to find pictorial elements which represent in a simple way the frequency-colour model. This seems to be feasible for nouns but is not easy for verbs and adjectives. An interesting method is the creation of a gallery of actors stored in a repository which are introduced into the scene by the learner to play small stories, for example the principal actors of mouse and elephant in Figure 1. An articulation error is shown by altering the maximum value of the contour, height and length are related to volume and rate, and omission errors can be visualized by different proportion of the body.

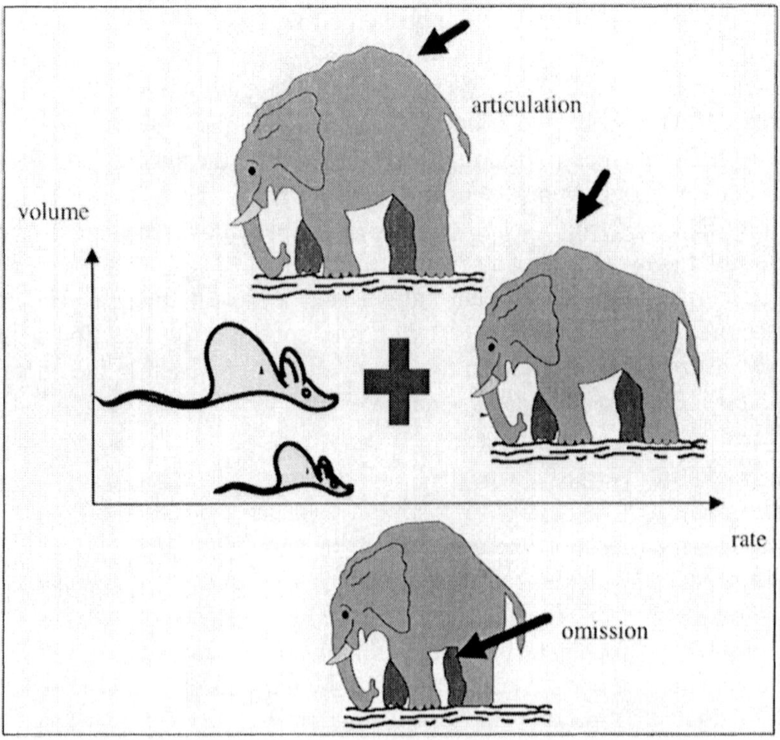

Figure 1. Error output by special representation of mouse and elephant

3.3 Evaluation and further work

To achieve a normative evaluation of Whisper, an interdisciplinary approach involving special schools, rehabilitation centers and self-help groups has been undertaken.

Whisper was presented at the REHA 99 exhibition in Düsseldorf and then evaluated during a six month period through special schools for hearing-impaired with the aid of all leading German rehabilitation centers. The usability of the English version was tested in association with the National Association for Deaf People in Dublin.

Most people admit to mistakes in volume, articulation, stress, and intonation and concede problems in understanding individual people, especially in group situations when voices are mixed. The test group plans to work with Whisper at home, to improve the pronunciation, to learn with sentences and stories, and to modify the repositories. In the future, we plan to enhance the speech error recognition, to analyze stress and intonation, to include a calibration routine, and to provide appropriate visual feedback. To meet the expectations of hearing-impaired people, we will include other dictionaries and an automatic generation of phonetic descriptions, implement further foreign language versions, develop benchmarks, and evaluate user progress. This work will be done in accordance with our modelling guidelines and will allow for evaluation of the proposed modelling process.

Many systems maintain classic but inappropriate interfaces and try to fix them in a way the impaired can interact with. We think we should try to develop totally new interfaces that allow people to develop mental structures or models of the world being represented by using the channels available (Baloian and Luther 2002) We may use the same devices but with a different logic behind them. Another important issue is how to motivate the user when there are no sound resources to produce special effects? How to trigger surprising effects, feelings or certain moods that are normally transmitted by a combination of high quality image sequences and sounds when one of these channels is not available? We think some answers can be provided by studying classic silent movies or radio theatre pieces.

4. CONCLUSION

In this article we have described a state of the art example in the field of systems for hearing-impaired people, especially from the point of view of transferring the real world into computer representations for this interest group. We have also explained a unified methodology for modelling the real world and have illustrated important tasks of defining error measures and adapted output formats in these systems.

A critical mass of educational systems have already been developed for disabled people, which allows some generalizations and recommendations to be made. The development of systems for people with disabilities should no longer appear as isolated handcrafted efforts; instead, the construction of these types of systems must be systematized. Recent advances in hardware and software developments indicate that the technological foundation for such systems has already been laid.

Acknowledgements

This research and common multi-media software development for the disabled (AudioDoom for blind people and Whisper for postlingually deaf people) are being carried out by the authors in the course of a current

German-Chilean project funded by the German ministry BMBF and the Universities of Duisburg and (Santiago de) Chile. We thank Dr. M. Mühlenbrock and Dr. D. Willett for their valuable suggestions and our collaborators Dr. W. Otten and C. Wans for their valuable support, discussing these ideas and contributing their research during the development and evaluation of Whisper.

REFERENCES

Alonso, F., de Antonio, A., Fuertes, J. L. and Montes, C. (1995) Teaching Communication Skills to Hearing-Impaired Children. *IEEE Multimedia*, Vol. 2, No. 4, pp. 55-67.

Baloian, N. and Luther, W. (2002) Visualization for the Mind's eye. To appear in *Dagstuhl SV Lecture Notes in Computer Science.* Springer, Berlin.

Breton, G., Bouville, C. and Pelé, D. (2001)FaceEngine. A 3D Facial Animation Engine for Real Time Applications. In *Proceedings of 2001 Web3D Symposium*, Paderborn, Germany, pp. 15-22.

Gräfe, J. (1998) *Visualisierung von Sprache und Erkennung sprechtypischer Parameter und ihre Veränderung bei Spätertaubten.* Diploma dissertation, GMU Duisburg.

Hobohm, K. (1993) *Verfahren zur Spektralanalyse und Mustergenerierung für die Realzeit-visualisierung gesprochener Sprache.* PhD dissertation, TU Berlin.

IBM (2000) *Speech viewer III*
[http://www-3.ibm.com/able/snsspv3.html]

ISAEUS (1997) *Speech Training for Deaf and Hearing-Impaired People*
[http://www.ercim.org/publication/Ercim_News/enw28/haton.html]

Lumbreras, M. and Sánchez, J. (1999) Interactive 3D Sound Hyperstories for Blind Children. *CHI '99*, Pittsburg PA, USA, pp. 318-325.

Visual Talker (1999)
[http://www.ed.gov/offices/OERI/SBIR/FY99/phase1/ph199t02.html]

Wans, C. (1998) An Interactive Multimedia Learning System for the Postlingually Deaf. *Proceedings of ITiCSE'98, 6th Annual Conference on the Teaching of Computing, 3rd Annual Conference on Integrating Technology into Computer Science Education*, ACM, New York, Oxford.

Wans, C. (1999) Computer-supported hearing exercises and speech training for hearing impaired and postlingually deaf. *Assistive Technology Research Series*, Vol 6 (1), IOS Press, pp. 564-568.

Willett, D. (2000) *Beiträge zur statistischen Modellierung und effizienten Dekodierung in der automatischen Spracherkennung*. PhD dissertation, GMU Duisburg, 2000.

Regional learning networks – building bridges between schools, university and community

Andreas Breiter
Telecommunications Research Group, University of Bremen, P.O. Box 330 440, 28334 Bremen, Germany

abreiter@informatik.uni-bremen.de

Key words: Digital Divide, Equity, Lifelong Learning/Education, Partnership

Abstract: The so-called "digital divide" seems to be a major social obstacle for the Information Society. Most experts agree that citizens will need competencies that go beyond the basic cultural skills. Where can our citizens acquire these competencies? The idea of life-long-learning illustrates a major problem of our educational institutions: they work separately, they only process the results of the preceding phase and there is a lack of interconnection. This paper tries to develop a concept of a regional education network that includes pre-school, K-12 and further education, public libraries and community centres as well as other educational institutions. Taking the path of the digital divide as a social and educational divide and focusing on the school as one major player in the regional network, the innovation process and the actors involved in it are highlighted and explored. Using action research in a project between schools, local community, private partners and the university, the concept of the regional network is illustrated. The problems encountered in the project are described and discussed in order to give a realistic overview and impression of an integrated education policy.

1. THE DIGITAL DIVIDE AS A CHALLENGE FOR EDUCATION POLICY

Most experts agree that the Internet will fundamentally change the way we live and work and the way we communicate in private and in public. Business processes within and between companies will change and so will public administration and political decision-making processes. Today, nobody is able to say with certainty what changes will take place but it is obvious that with increasing competition, organizations and individuals who understand the new digital media and are able to use it for their own purposes will gain certain advantages.

It is debatable whether the competent use of the various Internet services really is a cultural skill to be compared with reading and writing. Without doubt the World Wide Web as a complex, distributed information resource can only be used effectively with the competence formerly found in specialized jobs such as information brokers, librarians, documentation experts or journalists - which can now be carried out in do-it-yourself style. However, just as nobody necessarily becomes a good craftsman when a DIY store opens, the availability of websites and search engines does not mean that people will use them competently to reach their goals. Hence, teaching and learning media literacy is at the top of the agenda of education policy. Equipping schools and universities with computers and Internet connections is massively supported, leaving decreasing budgets for other aspects of educational provision.

However, schools only serve just over a bit more than 10 per cent of the overall population. Recently, a representative study found that nearly half of the German population are Internet 'objectors' (TSN-Emnid 2001). This notion suggests that these people had thoroughly investigated the Internet and had come to the conclusion that there is nothing relevant for them. Obviously, this is not the case: they have no experience and normally they will not buy a computer to find out the opposite. There are people who master the challenge and learn to use the new digital medium with more or less help from their personal or work environment. After all, about 40 per cent of Germans have Internet access and in other countries the proportion is even higher.

Closer examination of the socio-demographic distribution of Internet users reveals a divide between the groups in respect of their educational backgrounds. Ten times more people with a university degree are using the Internet than people without a high school degree. Therefore, those with a good education are expanding their leading position by using the new medium, while the disadvantaged are further falling behind. This can become a vicious circle because you can only learn to use the medium if you are able to master certain cultural skills, especially reading. This means that the digital divide does not create a new segregation, but it is increasing the existing educational differences. Among 80 million Germans, 16 million work at home, 4 million are unemployed and more than 20 million are older citizens. Beside initiatives like 'Schulen ans Netz' there are now initiatives in state institutions for special user groups, in public libraries, adult education centres, Community Technology Centres and youth centres as well as in commercial Internet cafés.

2. THE LEARNING EFFECTIVENESS PARADOX IN SCHOOLS

The ultimate goal of computer and Internet use in schools is better teaching and learning - usually measured by student achievement in standardized tests. According to studies on learning effectiveness in the last decade there is no hard evidence that the use of digital media for learning is more effective than traditional methods. Some studies found an increase in students' and teachers' motivation (Bertelsmann 1997), other studies noted ambivalent results in achievement tests (Wenglinsky 1998) depending on the way digital media were introduced into the curriculum (for an excellent overview see (Kulik 1994)).

Although there is a lack of empirical evidence, politicians, industry and foundations put billions of Euros into hard- and software, telecommunication infrastructure, curriculum development and teacher training. This can be called the 'learning effectiveness paradox'. It is derived from the well known 'productivity paradox' described in Information Systems literature - despite the heavy investment in IT in companies, productivity is not increasing or at least is hard to measure (Brynjolfsson and Hitt 1996).

What are the educational deductions from the 'learning effectiveness paradox'? If one argues economically, then any further investment in digital media in schools is wrong - the traditionalists argue for more teachers and less media. On the other hand, as we know from other organizations, media are experience goods that develop their value through usage and all organizations go through a learning process. In order to describe the interdependence of technology diffusion and organizational learning, different phase models have been proposed (Nolan 1993). What they have in common is that they outline a transition from an uncoordinated start phase to a targeted phase of steering and integration of technology into working processes and organizational structures (Figure 1). Technology has to be embedded in organizational processes again and again, and new structures and processes have to be created.

If we look into the process in schools and other educational institutions, this description highlights a lot of similarities. This combination of organizational learning and the integration of educational technologies was perfectly described by the study of RAND Co. for the U.S. Department of

Education which concluded: "Technology without reform is likely to have little value; widespread reform without technology is probably impossible" (Glennan and Melmed 1996).

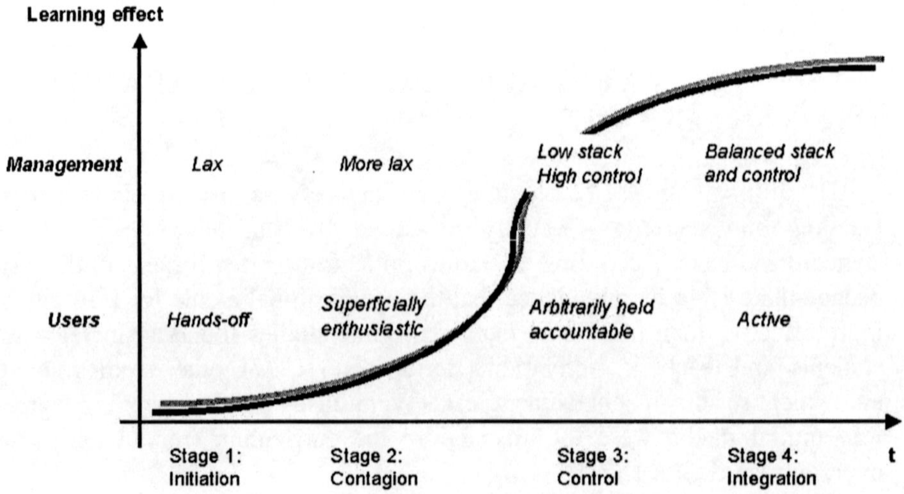

Figure 1. Development phases of IT use in organisations

After the eruption caused by the PISA study (Programme of International Student Assessment) the role of teaching and learning returned to the focus of politics, practice and research. This has considerable relevance for the use of digital media in schools. In an analysis, the North Central Regional Educational Laboratory (NCREL) described the ideal type of "engaged learning" (North Central 1997) and maintained it was built on the following basic principles:

- Children are engaged in authentic and multidisciplinary tasks.
- Assessments are based on students' performance of real tasks.
- Students participate in interactive modes of instruction.
- Students work collaboratively.
- Students are grouped heterogeneously.
- Students learn through exploration.
- Teachers are facilitators in learning.

One project to support authentic learning and provide a good combination of school-based activities with the local community to combat the digital divide has been set up in the State of Bremen.

3. THE WEBPOINTS

Using action research as a method is fairly new in information systems research. There are numerous definitions, however one of the most widely cited is that of Rapoport's. He defines the aims of action research "to contribute both to the practical concerns of people in an immediate problematic situation and to the goals of social science by joint collaboration within a mutually acceptable ethical framework" (Rapoport 1970). Action research focuses on collaborative aspects and possible ethical dilemmas that might result from its use. It is a valid research method in fields such as organizational development and education. However, in information systems, there seems to be increasing interest in action research although it was largely ignored for a long time (Baskerville and Wood-Harper 1998). In this paper, action research is used as an interpretative methodology to construct the basis of a formative evaluation, i.e. to learn from experience in the organization development process.

Most German Laender have enacted in their Education Laws that schools should open towards the local community. The school is part of public life and, as such, should play a major role in life-long-learning together with other education institutions. In Bremen, this gap between schools and their local communities might be partly filled by the project 'Webpoints'. The author was leader of a project team, managing the implementation process. This public-private-partnership between the State of Bremen and German Telecom has invested about 2 million Euros over two years (2000-2002) to build up Internet-Cafés in 25 schools in Bremen and Bremerhaven. The new computer labs provided in the schools can be used for teaching and learning in the morning hours, in the afternoon they are open to the public. The aim is to give schools the opportunity to establish partnerships with the local community via the medium of the computer. The key aspect of the concept is the support of external users by students of the schools - the 'webscouts'. They are given training in the basic technical skills and in instructing by university students. The webscouts give external users advice in how to find websites, how to install a personal e-mail account, how to use a spreadsheet or how to print out a text. This brings a double effect: university students in education get field experience and a new feeling in how to deal with kids. The webscouts develop important skills - technical know-how and especially the acquisition of social competencies in dealing with lay users.

The schools had to go through a difficult application procedure before they were accepted onto the project. This was new to them. An external committee at the University judged their applications according to several criteria:

- Quality of concept and IT-plan.

- Location (schools from disadvantaged regions with better chances).
- Quality of the room (free, separate, external access).
- Opening hours (5 hours a day from 2pm to 7pm, at least 15 hours per week of free access).
- Selection and support of webscouts.
- Co-ordination with parents association ('Schulverein') about the financial organization (payment for webscouts, ongoing costs).

The 25 winners had to sign a contract (another new experience) in which these criteria were laid down. After signing the schools received the following services and infrastructure resources from the project partners:

- The selected room was refurbished, painted and attractively renovated.
- A high value technical infrastructure (15 PCs, a laptop, peripherals, T1 Internet connection line and a LAN) was provided in each school.
- The webscouts were trained on 2-day training schemes and received a certificate after the successful completion of the course.
- The parents association received funding to pay the webscouts and other personnel to support the external users and finance ongoing costs.
- Co-ordination and support from the project management team at the University of Bremen.

Technical support is provided by a student organization. For the first two years, students of different faculties (but mainly from Computer Science) have helped schools with their technical problems. They have set up a help desk with a call center and a web-based FAQ-list. Additionally, they have worked on-site in schools. After several months of intrinsic motivation the district has taken over the financing of the project. Today, the project is a non-profit organization, is completely self-organized and employs 15 students and 5 technical experts. The School-Support-Service (S3) is working in more than 50 per cent of the 160 schools in Bremen. For the webpoints, they developed a reference installation for the complex case of double usage: teachers and students in the morning, external users in the afternoon. For students in computer science, this project offers more than a basis for living - directed by university lecturers, they build their own structures and processes, learn how to deal with difficult clients (teachers) and get additional technical know-how.

The first year of implementation of the webpoints revealed a lot of problems, especially in regard to construction sites and refurbishing and the co-operation with district officials about the network infrastructure. Additionally, some schools had difficulties in finding external partners to co-operate with. Schools with a long tradition of community integration experienced no problems - others did not even know whom to contact! On the other hand, after finishing the sites, the response from the local community was overwhelming. In many webpoints, senior citizens took the opportunity to learn how to use the Internet and lost some of their fears. This was the biggest advantage of the project had over other public Internet access points because the webscouts could give help. Effects of increased communication between the different age groups are difficult to measure, but site visits and qualitative interviews show impressive results. It appears both webscouts and external users have profited from each other. Apart from the very strong user group of senior citizens, the sites also offered additional courses:

- Reserved surfing hours for women, girls and families.
- Webscout courses for administrative personnel and cleaning staff.
- Theme days (Job-Fairs, video and music groups).
- Courses for disadvantaged groups (students, adults with special needs).
- Joint projects with community interest groups (city information systems, migration information system).

In summary, a threefold co-operation network could be created between:

- Schools and local community,
- Webscouts and external users,
- Schools and university (students, teachers, webscouts).

4. REGIONAL EDUCATION NETWORKS

In order to establish the webpoints as another form of public access points, it is necessary to embed them into the regional education infrastructure. The integration of computers as digital media requires complex coordination between all levels of the school system, making the separation between state and district affairs obsolete. The school district is the key regional player for this innovation process. This had already been stated in the late 70s during the large-scale evaluations of the new U.S. federal education policy reforms - "In short, the groundwork and planning for sustaining a change agent project had the early, active, and continued attention of school district managers" (Berman and McLaughlin 1978). Furthermore, no innovation in schools can be successfully implemented without the principal. The results of several studies in Germany (Prasse and Scholl 2001) confirm our early findings in respect of digital media that "The principal was the key to both implementation and continuation".

From the perspective of a district or a county that wants to offer all people the chance to learn to handle the new digital media, the strategic approach adopted in the school system is not effective. It neglects the fact that only 10 per cent of the population are in schools. In order to achieve economic and social goals for structural change from the industrial to the information society, the institutions of adult education are of equal importance. If life-long-learning is more than a shallow term in politics, media literacy has to be relevant in all institutions of the education system. To solve these questions, the institutional framework of the school system has to be reformed. Other public or private institutions have to be involved. This will only succeed if the political will (leadership) exists and concrete incentives are available. To date, there are no experiences with, or detailed studies of, these cross-institutional approaches. There is also a lack of evaluation criteria on how to measure the trans-organizational learning process. This situation is typical for innovations. In all larger institutions, several innovation processes are run in parallel but all at different phases. Implementation will take time. The webpoints project proved that the innovation is worthwhile.

REFERENCES

Bertelsmann-Foundation (Ed.) (1997) *Integrating Media into the Curriculum: Lessons from Athens Academy*, Verlag Bertelsmann Stiftung, Gütersloh.

Brynjolfsson, E. and Hitt, L. (1996) *Paradox Lost? Firm-level Evidence of the Returns of Information Systems Spending*. In Management Science, 42 (4), pp. 514-558.

Berman, P. and McLaughlin, M. (1978) *Federal programs supporting educational change*, Vol. VIII. Implementing and sustaining innovations, RAND Co., Santa Monica, CA.

Baskerville, R. and Wood-Harper, A. T (1998) Diversity in Information Systems Action Research Methods. *European Journal of Information Systems*, 7, pp. 90-107.

Glennan, T. K. & Melmed, A. A. (1996) *Fostering the Use of Educational Technology: Elements of a National Strategy*. RAND Corporation, Santa Monica, CA.

Kulik, J. A. (1994) Meta-analytic studies of findings on computer-based instruction. In *Technology Assessment in Education and Training*, E. L. Baker and H. E. O'Neill (eds.), Lawrence Erlbaum, Hillsdale, NJ. pp. 9-34.

North Central Regional Education Laboratory (1997) *Developing a Plan for Learning through Technology*, Illinois State Board of Education, Center for Learning Technologies, Illinois.

Nolan, R. L. (1993) *The Stages Theory: A Framework for IT Adoption and Organizational Learning*, Harvard Business School, Cambridge, Ma.

Prasse, D. and Scholl, W. (2001) Wie funktioniert die Interneteinführung an Schulen? In *Das Internet und die Schulen: bisherige Erfahrungen und Perspektiven für die Zukunft*, R. Groner and M. Dubi (eds.), Huber, Bern, pp. 63-83.

Rapoport, R.N. (1970) Three Dilemmas in Action Research. In *Human Relations*, (23:4), pp. 499-513.

TNS-Emnid (2001) *Der Verweigereratlas*. Initiative D21 und TNS Emnid, Basiserhebung.

Wenglinsky, H. (1998) *Does it compute? The relationship between educational technology and student achievement in Mathematics*. Policy Information Centre. Education Testing Service, Princeton, NJ.

Online knowledge communities: meeting places for continuing professional development

Sjoerd de Vries
Communication Studies, University of Twente, PO Box 217, NL-7500 AE Enschede, The Netherlands

vries@wmw.utwente.nl

Key words: Basic Concepts, Electronic Communication, Networks, Open Flexible Learning

Abstract: This paper describes the concept of Online Knowledge Communities (okc) as meeting places for continuing professional development (cpd). An okc is defined as a social network of members, who are online and are organized by making use of an online knowledge center. The okc has a particular group culture and members are involved in appropriate information processes to develop and exploit a specific knowledge domain. The author anticipates that the application of okc for cpd will have positive effects on the professionalism of the community members and on the development of knowledge assets in knowledge intensive organizations.

1. INTRODUCTION

Attractive online places for the continuing professional development of professionals, legitimate places, fully personalized, offering a wide range of knowledge services, flexible, highly interactive, and reliable - that has been the focus of our research into the development of successful online knowledge communities.

Our definition of a professional is a practitioner who possesses specialized knowledge and skills, often gained as a result of long and intensive academic preparation. They are characterized by or conform to the technical or ethical standards of a profession, operate with a high degree of autonomy and exhibit a courteous, conscientious, and generally businesslike manner in their workplace.

Continuing professional development (cpd) is needed for a number of reasons - the continuing development of professional competences, the rapid development of specialized knowledge and technologies and the increasingly flexible working contracts that force professionals into constant competition with fellow practitioners. The big issue for a professional is not how to become a practitioner, but rather how to stay one and develop further as a professional.

Current developments in Information and Communication Technologies (ICT), especially the developments associated with the Internet, have led to an enormous growth in online educational and training services for all educational levels, including professionals (e-learning). The main characteristic of these services is the management of courses. Educational institutions offer course management environments like Blackboard (www.blackboard.com) and WebCT (www.webct.com) to accommodate courses for a wide variety of professions. Such online courses hold great potential for the efficiency and effectiveness of online institutional-based education, mainly due to their flexible, interactive, multimedia-oriented, and personalized character. In general however, online courses still have a schedule, are time-framed, focus on a specific defined theoretical-based topic and are supply-driven.

Professionals are initiated into the traditions of communities of practitioners and the practice world they inhabit (Schön 1987). They seek just-in-time legitimate applicable knowledge, they need to reflect-in-action, to train themselves at the workplace, and be supported by social networks of fellow professionals. Clearly, online courses offered by educational institutions have a limited value for professionals. However, characteristics of online services like flexibility, interactivity, multimedia and personalization are essential for efficient and effective continuing professional development.

The concept we want to present here as an answer to cpd is 'online knowledge communities' (okc). We describe an okc as a social network of professionals aiming at the development and exploitation of a particular knowledge domain that is, to a certain extent, online (Vries et al 2001). Simply stated, we provide the communities of practitioners with electronic networked information and communication facilities and, based on these facilities, we broaden the scope to communities that develop and exploit particular knowledge. As Wenger and Snyder (2000) explain, 'a new organizational form is emerging that promises to complement existing structures and radically galvanize knowledge sharing, learning, and change'.

So, from our point of view, these networked facilities do not just support communities of practitioners but enable community forms that would not be possible without them.

2. CONTINUING PROFESSIONAL DEVELOPMENT

This concept has been around for a long time. We prefer the use of cpd in contrast to lifelong learning or continuing education because professional development occurs due to formal and informal learning in the course of practice - for instance by 'reflection-in-action' (Gardner 1978; Schön 1987). Formal and informal learning are synergystic and are vitally important for the development of professionals. So cpd is not only about learning or education - it is about a continuing process of professional development.

Cpd is in tune with 'modern' concepts like the learning organisation (Senge 1994), the knowledge creating company (Nonaka and Takeuchi 1995) and the theory that the continuing development of knowledge assets can secure the competitive advantage of organisations in the information economy (Boisot 1999). Boisot describes knowledge assets as stocks of knowledge from which services are expected to flow for a period of time, while others speak of organisation specific resources that are indispensable to create values for the organisation (Nonaka et al 2000). These authors try to concretise organisational knowledge by referring to 'stocks of knowledge' and 'specific resources' and they stress the significance of the assets for organisational success. These assets can be the professionals themselves, documents, methods and best practices. Effective cpd implies that the development of professionals secures the optimal development of knowledge assets, which in turn has a positive effect on professional services.

So, cpd is not only a must from the perspective of a professional, but it is a shared responsibility of the organisation (as the professional workplace) and the professional. We see cpd as a process of continuing development of knowledge, skills and attitudes of professionals, by means of formal and informal learning in the course of practice.

3. ONLINE KNOWLEDGE COMMUNITIES

We stated in the introduction that we consider okc as places for cpd, or to be more precise the places for continuing development of knowledge, skills and attitudes of professionals, by means of formal and informal learning in the course of practice. The question is, what are okc?

The concept of okc is based on the ideas behind virtual communities:

"Virtual communities are social aggregations that emerge from the Net when enough people carry on those public discussions long enough, with sufficient human feeling, to form webs of personal relationships in cyberspace." (Rheingold 2001)

So, the essence of a community is a stable, social aggregation based on communication - in this case on online communication. Here virtual refers to 'Internet-bound' instead of 'space-bound'. We prefer to speak of online communities instead of virtual communities. Virtual implies that it does not really exist and we want to consider existing, real, active communities. In general, online communities will have a specific mission, which aggregates 'the group of people'. The mission can be grounded in marketing (online marketing communities), business (online business communities), shared experiences (communities of practice) or specific interests (online health communities).

Online communities can be characterised on a number of dimensions. These dimensions reflect specific traits that describe the online community. Examples of dimensions based on organisation studies include: formalisation - the amount of written documentation; specialisation - the degree to which community tasks are subdivided; complexity - the number of sub activities or subsystems within the organisation; goals and strategy - the overall purpose and scope of the community; community technology - the actions and techniques used to change community inputs in outputs (Daft 1995). Besides the community technology dimension that means the community functions for the greater part online, we seek to identify the community strategic goals in the continuing development and exploitation of a particular knowledge domain. We interpret knowledge here as the set of knowledge assets of an okc, which are exploited by professional services.

In our research in the development of okc, we take a social network approach (Knoke 2001). Knoke mentions three basic assumptions of this approach:

- that the social structure of any complex system consists of stable patterns of repeated interactions connecting social actors to one another;
- that these social relationships are the primary explanatory units of analysis, rather than the characteristics of the individual actors;
- that the perceptions, attitudes and actions of organisational actors are shaped by the larger structural networks within which they are embedded, and in turn their collective behaviours can change these networks' structures.

We see an okc as a social network of members, who are online and are organized via an online knowledge center. They have a satisfactory or acceptable group culture and are involved in appropriate information processes to develop and exploit a certain knowledge domain (Figure 1).

At the heart of an okc is the professional network - chains of associations that link professionals, the members of the community, with common interests, problems and goals. In this network we distinguish various roles, which describe the expected behaviours, responsibilities and privileges of members. Examples of such roles are students, teachers, junior researchers, senior researchers and practitioners. Members may switch roles depending on the context. In Figure 1 we illustrate three examples of roles: practitioners, apprentices and experts.

The professionals in this network interact (in the sense of mutual or reciprocal action) in order to develop and exploit a particular knowledge domain and to develop and maintain social associations. Based on the work of Klein & Hirschheim (1993) who described the conceptual and philosophical foundations of information system development, we distinguish two dimensions of 'knowledge community interactions' - work and social interaction. Work represents the systematic struggle that members perform in response to inputs in order to achieve given ends, thus enabling the okc to develop and exploit knowledge and legitimatise its existence in relation to its environment. Typical work activities include experts who research, practitioners who develop themselves or coach others, and apprentices who learn or practice. Social interaction ideally involves all the relevant stakeholders of the okc. By utilising experiences and other information gained from that okc, as well as from social interaction with the environment, the stakeholders of social interaction should communicate to reach mutual understanding about the development and exploitation of knowledge and to change it accordingly. Examples of social interactions are meetings, discussions and chat sessions in which the work is discussed.

This professional network is organised, so an okc can be considered as an organisation although it may not necessarily be an organisation in the sense of a legal body. An organisation can be seen as a social entity that is goal-directed with deliberately structured activity systems and a permeable boundary (Daft 1995). An okc can be seen as an adhocracy, characterized by little formalisation, coordination by mutual adjustment, highly specialised tasks, a high level of formal education and training, a typical clan control and little hierarchy of authority in the sense of span of control.

An adequate (i.e. enough for the purpose) community culture is crucial if an adhocracy is to function. A culture can be described as the shared set of values, beliefs, behavioural expectations, and artifacts that comprise the way of life of a people (Renzetti and Curran 1998). Culture distinguishes one group of people, for instance the members of an okc, from another and gives members a sense of belonging - a feeling of 'we-ness'. The importance of a community culture is described by Daft (1995) who refers to an organisational culture as:

"it is culture that guides day-to-day working relationships and determines how people communicate within the organisation, what behaviour is acceptable or not acceptable, and how power and status is allocated."

Renzetti & Curran describe six cultural elements: symbols, language, values, norms, technologies and artifacts. When developing an okc cultural choices will be made with respect to these elements.

As we already explained community technology is a critical dimension allowing the community to function online. The function of these technologies can range from information distribution to highly developed collaboration and support platforms. We use the concept of an online knowledge centre as the core okc technology. An online knowledge centre is seen as a place for online interaction, offers the services needed for professional development and for the development and exploitation of knowledge. Online interaction, from a system point of view, can be described in many terms - sender(s), receiver(s), (a-)synchronic, media rich, noise or media formats. Examples of services are information access, information dissemination, communication, meeting, research, support, coaching, training and study and knowledge management.

The centre has to be usable i.e. easy to learn and easy to use. It works from the technological perspective by means of the Internet. Technically it is digital, networked, hypermedia and database-driven. Digital data means that all information (files, messages, chats, archives, discussions) is digitized, so it can be easily edited, maintained, transmitted, copied, and disseminated - all at a low cost by means of computers. Networked means that tasks and activities take place through a collection of interconnected networks of computers spanning the globe and running the Internet Protocol. As a result members of the group are flexible in place and time - performing tasks and activities jointly with other members. The concept of hypermedia refers to the integration of multimedia and hypertext. Multimedia can be described as an integration of text, graphics, still and moving images, sounds and any other medium where every information object can be represented, stored, transmitted and processed digitally. Importantly these objects can be very different, for instance a multimedia document, an online video communication or an online discussion. Hypertext is a text with links to other texts. So, hypermedia can be described as pieces of multimedia information with links to others. Hypermedia is attractive and flexible in use and is a particularly effective medium due to its media richness. The last technical characteristic mentioned is database-driven, implying that data is organised in records in one or more databases. This allows members to organise, access, and store data in a variety of formats. Members can be provided with information services that provide comprehensive information.

The last okc trait we want to mention is the appropriateness of information processes. These processes refer to the ways members carry out concrete tasks and activities. From the perspective of an okc, these tasks and activities concern the continuing development and exploitation of knowledge, skills and attitudes of professionals. Based on Boisot (1999) we distinguish three basic information processes - codification, abstraction and diffusion.

Codification is a process of giving form to phenomena or to experience. Abstraction is a process of discerning the structure that underlies the forms. Diffusion is a process of spreading information.

Boisot brings these three processes together in the I-space, which can be used to examine the flow properties of information within different agent groupings as a function of its degree of codification and abstraction. That means that it is possible to design typical information processes based on such flow properties. Examples of typical processes are to elicit, to train, to share, to inform and to publish. These typical processes have to be formalised, indicating their importance for the development and exploitation of knowledge.

We have described eleven okc traits, which are presented in Figure 1. These traits are interrelated. The social associations, the interactions and the community members represent the professional network. This network is organized as an adhocracy, has a typical community culture, and, in order to carry out concrete tasks and activities (information processes) makes use of an online knowledge centre. The strategic goals imply the development of knowledge assets exploited by professional services.

We mentioned that an okc functions for the greater part by means of the Internet. Naturally not all member activities - communication, cooperation or training - have to take place on the Internet. In many situations, if possible, face-to-face communication is preferred to online communication. This is not always possible, effective or efficient. However, the media richness of online communication will continue to develop, due to the growth in bandwidth of data.

People will become increasingly used to online communication which may exercise a positive effect on its quality. Despite these developments, a mixture of face-to-face and online communication will take place depending on member contexts.

We have described the concept of online knowledge communities, a social network of members, who are online, are organised and who make use of an online knowledge center. The okc have an adequate group culture and are involved in appropriate information processes to develop and exploit a certain knowledge domain. Just what do we expect from the use of okc for cpd?

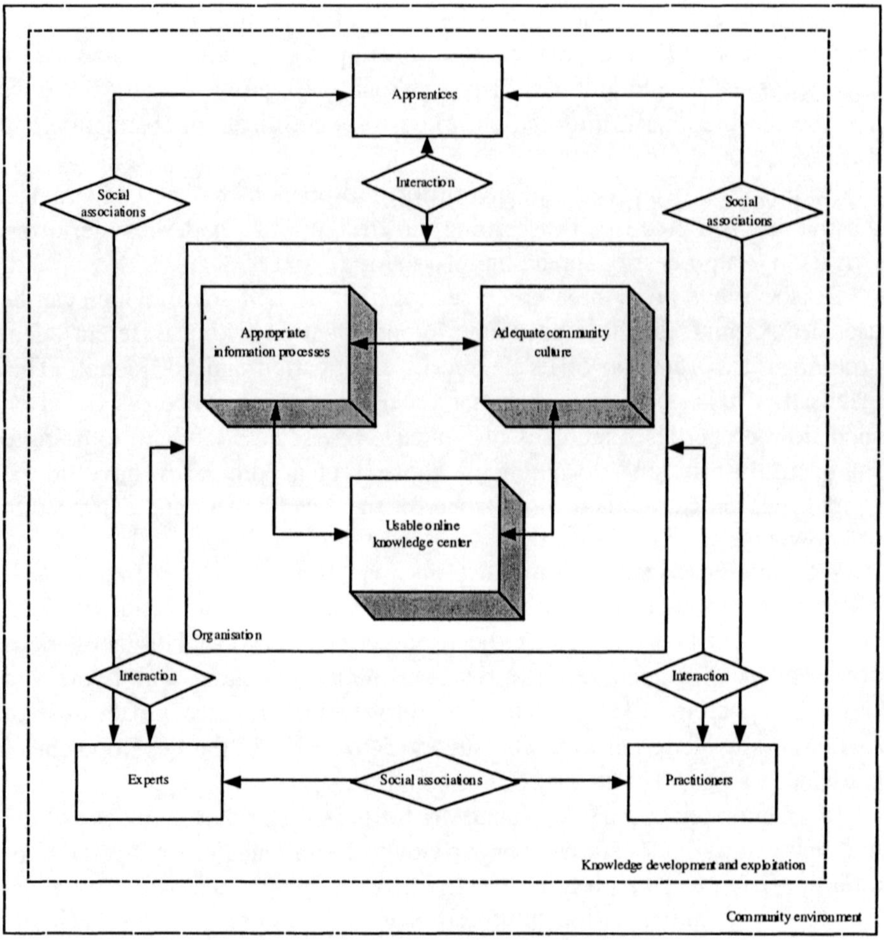

Figure 1. An online knowledge community

4. CONTINUING PROFESSIONAL DEVELOPMENT IN ONLINE KNOWLEDGE COMMUNITIES

As stated, we see cpd as a process of continuing development of knowledge, skills and attitudes of professionals, by means of formal and informal learning in the course of practice. The use of okc for cpd implies that an okc has to support this process. Becoming a member of such a community, a professional will have a place for cpd that gives:

- individualised, flexible and easy access to a coherent and actual knowledge domain;
- a range of opportunities to interact with like-minded persons;
- a range of opportunities to develop and exploit the knowledge domain, by for instance applying knowledge, learn from it, guide others, disseminate ideas and results or doing research;

in a professional network.

But the question to be answered is: what do we expect from this use? Given the discussion, our premise is that the membership of professionals of an okc has positive effects on:

- the professionalization of the professionals, expressed in competences like knowledge, skills, experiences and attitude;
- the development of the knowledge assets, expressed in a growth and elaboration of the professional knowledge, applicability of knowledge, legitimacy of knowledge.

However, the organisation of the okc has to be adequate, the professional network has to be well developed and has to ensure the continuing development and exploitation of the knowledge domain. Traits of okc can be seen as factors that influence the development of the professional network in an okc.

Insight into these factors, which can be held responsible for the development of okc as meeting places for cpd, helps us to identify the critical success factors of okc's and cpd, to identify the relationships between these factors, and to intervene in the development of okc.

5. CONCLUSION

Clearly, we expect that the use of okc for cpd will have positive effects on the quality of continuing professional development. Based on our concept, a number of okc are developed, for different knowledge domains (See for instance http://www.concourse.nl). We will research the development based on the described development factors and the results will give us insight into the effects.

REFERENCES

Boisot, M. H. (1999) *Knowledge Assets. Securing competive advantage in the information economy*. Oxford University Press, Oxford.

Daft, R. L. (1995) *Organization Theory and Design*. West Publishing Company, St. Paul.

Gardner, R. (1978) *Policy on continuing education: A report with recommendations for action*. The York Centre, University of York, York.

Klein, H. K., and Hirschheim, R. (1993) The application of neohumanist principles in information systems development. In *Human, organizational and social dimensions of information systems development*, D. Avison, T. E. Kendall, and J. I. DeGross (eds.), Elsevier, Amsterdam. pp. 263-280.

Knoke, D. (2001) *Changing organizations. Business networks in the new political economy*. Westview Press, Boulder.

Nonaka, I. and Takeuchi, H. (1995) *The Knowledge Creating Company. How Japanese Companies Create the Dynamics of Innovation*. Oxford University Press, New York.

Nonaka, I., Toyoma, R. and Konno, N. (2000) *SECI, BA and Leadership: A unified Model of Dynamic Knowledge Creation*. Long Range Planning, 33, 5-34.

Renzetti, C. M. Curran, D. J. (1998). *Living Sociology*. Allyn and Bacon, Boston.

Rheingold, H. (2001 july) Virtual community services.
[http://www.rheingold.com.]

Schön, D. (1987) *Educating the reflective practitioner*. Jossey-Bass Publishers, San Francisco.

Senge, P. M. (1994) *The fifth discipline. The art and practice of the learning organization*. Currency Doubleday, New York.

Vries, S. A. de, Roossink, L. L. and Moonen, J. C. M. M. (2001) *De ontwikkeling van online leercommunities in het studiehuis*. NWO 411-211-04. Universiteit Twente, Enschede.

Wenger, E. and Snyder, W. M. (2000) *Communities of practice: The organizational Frontier*. Harvard Business Review, 78(1), pp.139-145.

Distribution of Internet community knowledge based on traditional Communication Media

J. Felix Hampe, Silke Schönert, Christian Dietze and Nhiem Lu
Institute for IS Research, University of Koblenz-Landau, Rheinau 1, D-56075 Koblenz, Germany

{hampe|schoen}@uni-koblenz.de

Key words: ICT, Knowledge Management, Media Education, Society

Abstract: In recent years the creation of Internet-based knowledge has become increasingly significant. However, with regard to the influence and control of knowledge management processes, knowledge communities indicate that there are specific problems in creating and distributing knowledge. People without Internet access are excluded from this knowledge dissemination. Communication Centre Initiative Rheinland-Pfalz (CCIRP) - a co-operative project involving various ministries and the University of Koblenz-Landau - was set up to address these problems. It has shown how concepts of knowledge generation in knowledge communities (e.g. cc-expert) can be realised and how this knowledge can be distributed using traditional media.

1. INTRODUCTION

Recent years have seen a dramatic increase in the amount of information stored in electronic format and available on the Internet. It has been estimated that the amount of information in the world doubles every 20 months. So-called knowledge communities are major supporters of the generation and distribution of knowledge. However, although many are able to participate in this knowledge dissemination others are left out.

Based on information supplied by the Federal Statistical Office of Germany, in 1998 9.4% of German households used modems for data exchange and 7.9 % had Internet access. By 2000 modems were available in 15 % of the households and Internet access in 17.4 % of them (Statistics Bundesamt 2000).

The goal is to integrate persons without Internet access into the knowledge dissemination process by using traditional communication media like telephone or fax. This paper describes an approach for developing knowledge in a knowledge community and distributing this knowledge by heterogeneous media.

2. KNOWLEDGE GENERATION AND KNOWLEDGE DISTRIBUTION

Knowledge presents a state that can be attained through processes of knowledge acquisition. The acquisition of knowledge is normally demarcated as learning, which can be divided into three fundamental phases in each individual - (1) understanding, (2) storing of knowledge, (3) recollection of knowledge (Steiner 1992).

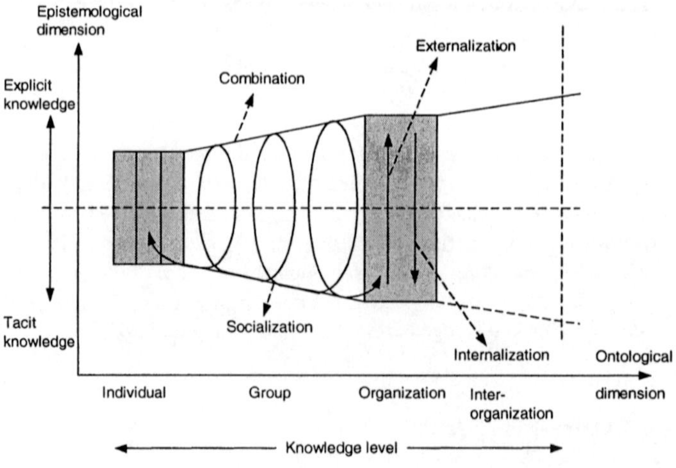

Figure 1. The spiral of knowledge

Individuals, artificially created media such as books and information and communication systems, groups, élites or whole organisations are seen as carriers of knowledge (Probst and Büchel 1994). Nonaka and Takeuchi developed the spiral of knowledge generation, which describes the process of knowledge transfer from individuals to teams and eventually the whole organisation (Figure 1). According to their model knowledge is created through repeated conversion activities of implicit and explicit knowledge. Implicit knowledge such as shared mental models or technical know-how is acquired non-verbally by imitating or watching experts. Explicit knowledge is available in the form of metaphors, analogies and models that are often presented verbally. The main forms of knowledge conversion are socialisation, internalisation, externalisation and combination (Nonaka and Takeuchi 1995).

2.1 Knowledge generation in communities

The Internet has become a so-called social medium, where people meet and communicate. People who meet frequently and engage in discussions and information exchanges based on an Internet platform are known as an Internet community or virtual community. A virtual community consists of an Internet user group who have a common interest or hobby. Virtual community members share ideas, experiences and information in the context of the specific community theme.

The information is shared in discussion forums, bulletin boards or chat systems. A community organiser can suggest discussion themes and community members create contributions or reply to contributions of other members (Shaw 2000).

On the Internet different types of communities exist (Schmidt 2000). One important type is the knowledge community. In this case knowledge is regarded as a product of communication between people. Knowledge Communities are Internet communities who's goal is to share knowledge (Schmidt 2000). All participants can offer and search for knowledge. It is important to create transparency about which knowledge exists in the community. Knowledge communities improve virtual knowledge exchange and enable the creation of so-called knowledge markets where users can participate by offering knowledge, by searching knowledge or act as a knowledge organiser - even a knowledge manager.

The intention of a knowledge community is to generate group knowledge. On the one hand this is achieved through the contributions of the members. On the other hand it is possible to activate community members by offering specific questions and discussion themes. The aggregation of a couple of contributions to a specific theme may be helpful to give a general idea of a topic. As well as the participation of interested persons it is intended that known experts participate in knowledge communities. This can be achieved by specifying expert consulting hours, creating expert discussion boards and encouraging expert contributions.

Creating a goal-oriented structure to community knowledge is a great challenge. One approach to making knowledge persistent is to build FAQ's and theme-specific structured documents. Another approach is to create knowledge networks. A knowledge network consists of knowledge objects and the relationships or links that exist between them. The addition of new relationships can be regarded as the creation of new knowledge. The central idea is to organise knowledge according to the concept of mind maps (Figure 2).

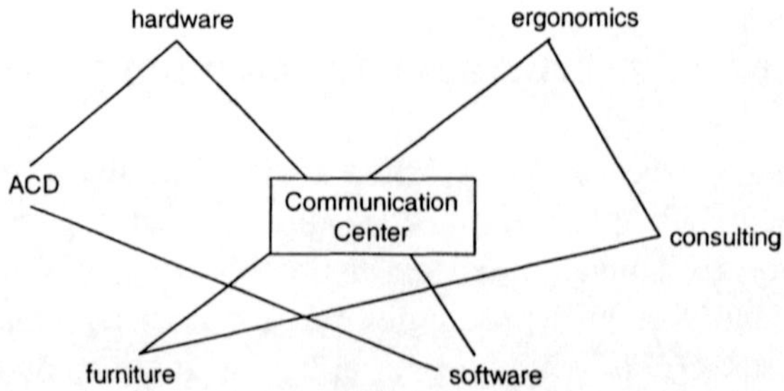

Figure 2. Example of networked knowledge

These provide a means of illustrating all relevant information for a specific topic in a graphical manner that mirrors theories about the way in which our mind works.

2.2 Knowledge distribution depending on communication media

There is a great amount of knowledge available on the Internet. One challenge is to offer access to this knowledge by using different traditional communication media. Experts may also be involved through this use of different media. An advanced circle of knowledge distribution is created (Figure 3).

This advanced or extended circle of knowledge distribution is determined by two aspects:

- which communication media (Table 1) is used to integrate the user-group?
- whether the integration can be achieved with high availability (7 * 24) and with low costs?

Table 1. Communication Media

Type	Communication medium
Voice	Telephone, Mobile phone, Radio contact
Graphic and text	Fax, Telex, Chat, WAP, e-mail, sms
Voice and graphic	Internet, Conferencing systems

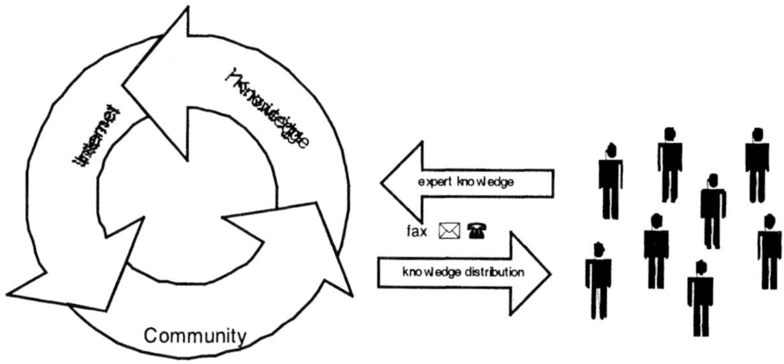

Figure 3. Extended circle of knowledge distribution in knowledge communities

Two traditional communications media are most appropriate for achieving integration of user-groups - telephone and fax. Telephone is still a favourite media for communication. The idea is to realise access to knowledge from any telephone, anywhere, at anytime. To fulfill the low cost criteria it is not appropriate to disseminate the knowledge using a person but rather by using speech technologies (Harbush et al 2001). Fax is also a favourite media. It is easy to handle, highly available and has low investment and low variable costs.

3. CCIRP

The Communication Center Initiative of Rheinland-Pfalz is a cooperative project between the ministry of work, social, family and health; the ministry of economy, traffic, agriculture and viticulture; and the university of Koblenz-Landau. Communication centres are call centres with a high level of media integration (Hampe and Schönert 1997).

The goal of the communication center project is to conceive an initiative to support the attractiveness of Rheinland-Pfalz for potential investors and thus create new jobs. With this in mind a project with four components was initiated. One component focuses on virtual communities. For the project a virtual community was built - known as cc-expert.

3.1 Virtual community

The virtual community cc-expert is a platform where people who are interested in the communication centre can meet, share information and consult with experts frequently (Dietze 2001). It is accessed at http://www.cc-exppert.de and consists of the following functions:

- cc-news (news about the communication centre, conferences, newsletter and cc-expert updates);
- cc-know-how (glossary, general information, communication centre locations, competitors, regulations and laws, communication centre sponsorship and communication centre technology);
- cc-personnel (their qualifications and education, comparison of wages in the communication centre, virtual employment agency);
- cc-market (offering market analysis and trends, information on sellers, consultants and education institutes);
- cc-calendar (communication centre events such as conferences, exhibitions, workshops);
- cc-contacts (contact to experts);
- cc-links.

The knowledge stored in cc-expert is organised as a knowledge network that consists of objects and the relationships between them.

The knowledge can be easily retrieved, either by actively searching through stored documents or by using certain filtering mechanisms. One of the latter is based on profiling. As a consequence, certain types of information are automatically sent to the user (push-mechanism) while others need to be actively sought by the user (pull-mechanism). The push mechanism can be effected using different communication media, for example new information objects which fit the user's profile will be sent automatically to the user (by phone or by fax). The technical platform used in the CCIRP reference lab is described in the following section.

3.2 Reference laboratory

In the project context a reference laboratory was designed and built at the University of Koblenz-Landau. The intention is to analyze how information can be disseminated using different types of communication media. This laboratory integrates Internet, phone and fax technologies (Figure 4).

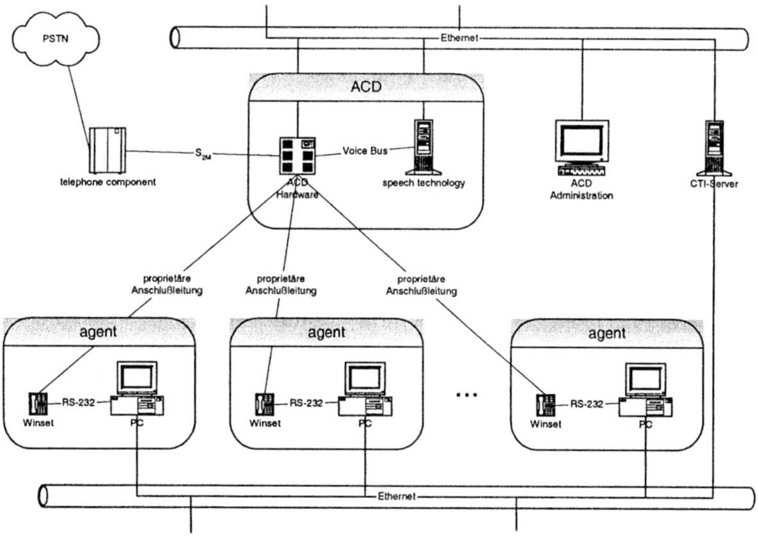

Figure 4. Structure of the Reference Laboratory

Knowledge dissemination is realized with heterogeneous communication media. These are combined in several ways. One possible scenario is that a potential investor calls the communication centre (by agent or voice system) to get information about call centre locations in the area. This investor is especially interested in financial support and contact persons. Via a speech technology based voice system he can choose between different options to:

- get an Internet address of an Internet page which contains the required information;
- get the information immediately via voice system;
- receive a fax message immediately;
- receive a letter by mail;
- get a call-back from a live agent.

After choosing the appropriate option the user can ask for further information. The system offers related information objects so the caller is able to navigate through the knowledge network.

4. CONCLUSION

We have presented an outline of a technology to generate and distribute knowledge. Our approach is intended to analyse how user-groups without Internet access can be integrated into knowledge communities. On the one hand the knowledge must be stored in an appropriate way, so it can be easily retrieved by different media, and on the other hand a comfortable platform offers different media interfaces. The first experimental uses have offered us various insights into the potential benefits. It remains to be seen, however, whether people who do not have Internet access will accept the system. Certainly the long term operation of cc-expert will lead to further requirements which arise from uses in daily life.

REFERENCES

Dietze, C., Hampe, J.F. and Schönert, S. (2001) VC3 – Konzeption und prototypische Realisierung einer Virtual Community im Communication. In *Virtuelle Organisation und Neue Medien 2001*, Engelien, M. and Homann, J., Josef Eul Verlag, pp. 307-327.

Harbusch, K., Knapp, M. and Laumann, C. (2001) Modelling User-Initiative in an automatic help desk system. In *Proceedings of Workshop on Natural Language Processing and Neural Networks (NLPNN2001)*. Natural Language Processing Pacific Rim Symposium, Tokyo, Japan.

Hampe, J. F. and Schönert, S. (1997) Call Center. In *Wirtschaftsinformatik* 39, 2, pp. 173-176.

Statistisches Bundesamt [http://www.destatis.de/presse/deutsch/pm2000/p3400024.htm]

Schmidt, M. P. (2000) *Knowledge Communities*. Addison-Wesley.

Shaw, M. (2000) *Handbook on electronic commerce*. Springer, Berlin.

Steiner, G. (1992) Lerntheorien. In *Handwörterbuch des Personalwesens*, E. Gaugler and W. Weber (eds.), Poeschel, Auflage, Stuttgart.

Probst, G. J. B. and Büchel, B. S. T. (1994) *Organisationales Lernen - Wettbewerbsvorteil der Zukunft*. Gabler, Wiesbaden

Nonaka, I. and Takeuchi, H. (1995) *The knowledge-Creating Company*. Oxford University Press, Oxford, New York.

Taking the best from real teaching environments

Iolanda Bueno de Camargo Cortelazzo
Universidade Tuiuti do Paraná, Rua João Azolin, 435 casa 13, 82015-040 Curitiba, Paraná, Brasil

iolanda.cortelazzo@utp.br

Key words: Collaborative Learning, ICT, Learning Environments

Abstract: This paper offers a description of the scenarios I have faced and the approaches I have used during my teaching experience. In particular I have used ICT to mediate knowledge construction and critical thinking skills development with under-graduate students in different technological environments in a number of private Higher Education institutions. This work has focused on the use of ICT as a collaborative learning and teaching tool. The reactions of students in the institutions have been surprisingly varied and the paper explores these reactions and the outcomes of the use of ICT.

1. INTRODUCTION

In many countries in Latin America and in other parts of the Western world, several management problems are created for teachers when there are too many students in Higher Education classrooms. Too many can mean up to eighty or ninety students in a single class under the responsibility of a single subject-specific teacher. Teachers understandably find it difficult to manage their classes and carry out cooperative and collaborative learning with their students

Most of the private colleges and some private universities have to deal with the economic difficulties faced by many of their students who belong to lower levels of the society. To maintain their business these institutions run most economically with large classes.

Educational environments, technologies and human resources vary markedly from institution to institution. Some offer very good working conditions and possibilities of continuing self-development for their staff, as well as creating good learning environments. However, most Higher Education institutions in Latin America employ hourly-paid teachers. Most of these teachers work in many different institutions in order to survive. That is a real concern to course coordinators and college deans because these teachers have neither the time nor the resources to invest in their on-going academic development and research.

I have been working in teacher development and running graduation courses since 1972, but my classes never had more than 30 students until 2000. In 2001, I started teaching under-graduation programmes with classes varying from 40 to 80 students in a single room. Many people say we can not have good quality in such large classes but teachers should try to work in a way that allows students to acquire and construct knowledge as well as skills. If we use ICT in an appropriate way we can develop our students' critical-thinking skills and their creativity.

2. TEACHING MEDIATED BY ICT - A WORKING SCENARIO

My main approach is to try and develop work that stimulates student co-operation and collaboration alongside their genuine will to learn (Cortelazzo 2000). Although the institutions where I have taught are private institutions, the students come from low medium classes and have low incomes. Consequently our mission is to ensure they get the most from the payments they make because these payments represent considerable sacrifice for themselves and for their families. They also represent a major investment in what they hope will be a better lifestyle.

Institution A has an up-to-date technological environment. There are three networked laboratories and every class is connected to the NT Windows-based network and to the Internet. Every student has an e-mail address and, through the college networking system, students have on-line access to the institution's facilities. Teachers can use the mimeo system, have all the class work prepared in advance and uploaded to appropriate folders at the college and save their electronic board class-work in their own folders.

Although the students soon mastered the operation of the system and used the network to communicate and do research, it was necessary to develop and implement a two-year programme to help most of the teachers use the ICT environment. Even so, their main use of the technological environment has been to respond to the administrative demands of the courses. Only a few use it pedagogically. As the pedagogical coordinators have realized this situation, they have offered some pedagogical workshops not only to stimulate teachers but also to offer them guidelines and suggestions to help them introduce ITC into their daily teaching routine.

Institution B is a traditional institution trying to choose the best technological solution for its specific needs. In the classrooms, on the campus I work at, we only have whiteboards and pens. If we want, or need, to use TV, video, overhead projector or the single forty-place stand-alone-computer laboratory, we need to assign our requirements a week in advance. Only the Course Coordinator's computer and three computers in the library have Internet access. About 30% of the students have free web-mail access or a paid Internet account of their own. Most of the teachers do not use Internet facilities, they simply use computers as typing machines and the ICT they are used to are books, chalk and boards. Few use video and television to support their class-work. Only very few access the Internet either for communication or for research. In fact, most of them do not carry out research. Consequently, their students are on a similar path. Many of the students have already been involved in teaching and their acquaintance with ICT is as narrow as that of their teachers. This situation does not seem to change even if the mastery and use of ICT is modelled to them at Teacher College throughout the subject-specific course *ITC in Education.*

In August 2001 I started working in another institution - institution C - a university with over forty different teaching courses. I have been teaching both on the Graduate Programme in Education and on the under-graduate programme at the Teacher College. The technological environment is more sophisticated and there is more technology available than at B. It is, however, poorer than at A.

There are many laboratories and all the teachers and students have an e-mail address (although many do not know they have it). Most of the teachers neither use computers as teaching tools nor access Internet as users - they just access it because they have to register the students' grades via the Internet. On the other hand this institution has a dynamic web site. Most of the news or information to teachers and students is delivered by e-mail or communicated on their homepage. They have an Intranet available for teachers and students. They also have a Department for Teacher Development that offers some workshops and lectures.

My research in this university is related to the use that former students of Pedagogy (who graduated from 1998 to 2001) make of ICT in their current employment. This research has generated information to help support my own teaching, general ICT teaching and the use of ICT in the Teacher College programmes.

3. TEACHING MEDIATED BY ICT - AN APPROACH TO TECHNOLOGY MEDIATING TEACHING AND LEARNING

I taught *New Technologies and Education*, the *Multimedia Laboratory* and *Distance Education* in the Teacher College (Institution A) and *History of Computing* and *Technology and Society* in the Computing Science Course at Institution B from February to June 2001. Since August 2001 I have been teaching *ICT applied to Education* at Institution C. Although the subjects are different, they have in common the objective of making students think about ICT and their application in their specific professional area.

Teaching these different subjects allows me to experience different points of views from technophiles to people who do not wish to work with machines - not only students, but also teachers and researchers.

Students are required to think about the concepts of education (teaching and learning as two sides of the same coin), communication and technology. Extensive class discussion helps students to consider and clarify their understanding of these concepts.

Teacher and students come to a common understanding of the work of the course. ICT is defined as any way that generates, stores, processes, distributes and reproduces information applied to solve any problem or create better life conditions. Paper, files, magnetic tapes and optical discs are means of storing information. Xerox machines, overhead projectors, slide projectors and data displays are means of reproducing information. Computers and robots are means of processing information. Fax, telephone, newspaper, magazines, mail, television, telex and computer networks (e.g. the Internet) are means of distributing information. On the other hand, students suggested that we, the teachers, are the first known effective 'ICT' at school, as we generate, store, process, distribute and reproduce not only information but knowledge with our students.

The idea of collaboration is also clarified. Collaboration is the basis for a consistent and productive partnership and both are essential to the completion of a project where joint construction is expected. In school education collaboration is also fundamental and must be present among students and between teacher and students. The next step with students is to state what collaboration is and clarify the its rules. Every student must make a meaningful contribution to the project according to his/her competence. Everyone has long-term objectives which must be inter-related to the collective objectives of their group. All participants need to bring their own values, skills and competence to contribute to an effectively functioning project. Fundamentally everyone must know exactly what their role and responsibilities are, what these limits allow, and the benefits or prejudices inherent to these functions and responsibilities. It is not necessary for everybody to be friendly but it is important they respect each other and work for the development of the group. It is also important to consider counter arguments, respect each other's thoughts, incorporate them when necessary and recognize that one's own convictions are not appropriate at that moment. Information must be shared, discussed, analyzed and processed by the whole group to transform it into enriching knowledge. From divergence and argumentation in a conscious dialectic may result rich and meaningful products for each of the group participants. Once the participants build up wide connections, develop interconnected work forms and share information as well as specific knowledge, they become teachers and students of each other.

Communication between partners is essential in collaborative projects. It can happen in different ways and need not necessarily be verbal but it must be continuous. Personal constructions may be enriched with other contributions, integration occurs, the interdependence is maintained and the process is always receiving feedback. Students are required to build up a 'sharing space' as a real test for collaborative action. ICT are necessary in this collaborative project since they allow teacher and students to develop mutual collaborations even if they are far apart. Telephone, fax and mail services like regular mail or Federal Express have being used for a long time. Electronic networks have expanded the possibilities creating a meaningful cyberspace characterised by collaboration.

Students are required to examine and consider these rules and reflect on the importance that socially shared knowledge holds for learning. Interactive collaborations may contribute to the shared knowledge construction and inter-subjectivity development (known as the capacity to have friendship and emotion with others). They set a shared reference regarding objects and external circumstances (Crook 1996).

Thus, in each course new collectives are formed, they collaborate mutually and construct new knowledge, aggregating value to the existing knowledge, weaving new relationship webs and new relations between different subjects, consolidating a collective, cooperative and collaborative intelligence.

All the courses have specific objectives and the approaches to each one are inevitably different but the position as regards knowledge acquisition and construction must be the same. Teacher College students will deal with learners (pupils) in educational environments, and Computing Science students will deal with learners (users, consumers, clients) in real world environments. All will be required to be competent, efficacious and effective in their field. They will have to know how to make decisions, provide solutions and be creative in their specific professional action. Therefore they have to be developed and trained to be able to make up their mind at the right moment, to provide the best solution, to be creative and also to be cooperative and collaborative. That is why they have to learn how ICT may or may not mediate their actions and preview possible problems and/or dangers these technologies may bring. They have to experience ICT as collaborative tools which can enhance their actions and promote their development as human beings. They have to reflect on their use of ICT and on the positive and negative consequences of its use.

The reactions I have experienced to the use of ICT in the three institutions have been surprisingly different. These experiences demand reflection from teacher trainers and College Coordinators.

The Scientific Computing students were more ready to study and work with ICT and to think about their use and their role than the Teacher College students. Many of the latter have already been working as teachers or trainees in elementary schools or kindergarten. They are resistant to innovation. They enjoy being involved in practical, hands-on work but do not like to write or read. Many are there just to get a certificate.

In all the three institutions students were required to read extensively. They were guided to summarise and make reviews of their writing. Writing in ink and on the computer, drawing with crayons, drawing on the computer, photographing, scanning photos and inserting them in MSWord texts were some of the required activities. Although the subject programme in the less technological Institute required computer activities - the computer laboratory was available only once a month. Students were stimulated to use other ICT.

They created projects preparing radio programmes recorded in audio; they videotaped activities and presented them on television, they prepared slides with presentation software and showed them to their peers as overhead projector transparencies.

Although the change in their behaviour and their beliefs was a slow process, it is encouraging to see it happen. Some times students are not aware they have changed - sometimes their attitudes and posture change a little faster than their consciousness of it.

It was hard work, but they started to realize how important it is to experience ICT, discuss its use, discover their unique uses and integrate them in Education. They have also realized that ICT will not change education nor solve the problem of quality in teaching. Bad teaching may become worse if ICT is introduced inappropriately.

The students I currently work with enrolled in June 2001 and started the Pedagogy course in August. There are 50 students and we should go to the computer laboratory for every class. However, we go twice a month because they are first required to read about technology and education, collaborative work and team work and then they are involved in class discussions. Finally they go to laboratory where they perform tasks, conduct research, write about their findings, create presentations and communicate with their peers and other teachers from different universities in other cities. They write a lot more, they discuss much more and they like what they do because they know that their work is not going to lie forgotten in a drawer. It may be published on the school bulletin board or on the WWW. As they can read their colleagues' or other students' work - theirs will be read too.

4. TEACHING MEDIATED BY ICT - OUTCOMES

When these students use the word processor, they write much more and with fewer grammatical mistakes. They are asked to follow three steps which are very difficult to follow when you write with pencil or ink - create, revise and edit. They do this collaboratively and enjoy it. Some are not enthusiastic about computers but are stimulated to use other technology.

Students from the three institutions are communicating with each other on the Internet. As they study different subjects, live in different communities and belong to different cultures, they have different points of view. These can enrich their exchanges. They discuss not only the use of ICT in education but also how they can help to improve humanity and be involved in the fight for peace in the world. Their participation in electronic forums increases as they realise their doubts, concerns, questions and comments are read and considered by many other people besides their teachers.

All students were required to finish the school year in June 2002 with some communication products produced and reflected upon during their course. These should include articles, photographs, audio and videotapes, digital texts and personal home-pages designed and created by themselves, individually and in work teams. These will be used in education and in their everyday life.

They have gradually realized that ICT offers an interesting media for them to promote good quality education, respect and solidarity, through individual and collective collaborative projects. They have learnt that communication and collaboration are the key to help change society, to enhance human relations and to make the world a better place in which to live.

Teachers and students are part of a cyberspace where different attitudes, beliefs and thoughts contribute to create a new scenario for a more meaningful and peaceful future for humanity.

ICT helps teachers at university, teacher trainers and human resource managers to work with different knowledge, different competencies and different skills. They help them to integrate both cognition and emotion to produce the best teachers and students and, above all, the best human beings.

Teachers cannot change attitudes, behaviours and beliefs in a short time. However, it is very rewarding to see students becoming involved in the projects and working together collaboratively. It is also impressive to realize how much teachers learn from them and how much we change and grow as professionals and human beings.

REFERENCES

Cortelazzo, I. (2000) Collaboration, Teamwork and the Technologies of Communication: Relationships of Nearness in Graduation Courses, Doctoral thesis, Faculdade de Educação da Universidade de São Paulo, Brasil.

Crook, C. (1996) *Computers and the Collaborative Experience of Learning*, Routledge, New York.

A Role-Based Adaptive CSCL Environment for Intensive Hands-on Teaching and Learning under Rigid Time Constraints

Horst F. Wedde, Frank Thorsten Breuer and Muddassar Farooq
Informatik III, University of Dortmund, 44221 Dortmund, Germany

{wedde, breuer, farooq}@ls3.cs.uni-dortmund.de

Key words: Learning Environments, E-learning, Didactics of Computing, Open Learning Systems, Role-based communication, Teaching methods

Abstract: Computer-supported Collaborative Learning (CSCL) is a widely accepted group-teaching and learning methodology. It focuses on communication and collaboration aspects among learning entities, between instructors and students. The contribution of this paper is twofold. Firstly we introduce a novel, role-based concept of collaborative learning where both instructors and students/student groups operate in combined teaching and learning activities while acting mainly autonomously. Secondly the beneficial learning effect of these methods (for students and instructors) is greatly enhanced by our new approach of reducing the course material to the essential underlying problems. In addition we shape the multimedia-based CSCL methodology according to the nature and character of the problem world of the course subject area. We apply our method to a classical area of computer operating systems (storage management) illustrated by the example of the famous 'Dining Philosophers' Problem'. The benefits of our approach are of particular use for senior-level undergraduate courses, for Life-Long Learning or for Distance Teaching and Learning programmes. In many of these areas intensive hands-on experience and competence has to be achieved under rigid time or learning environment constraints. In traditional classroom teaching these constraints would typically only allow for a superficial acquaintance with the subject areas.

1. INTRODUCTION

During the past few years, European universities have redefined their education programmes by introducing B.Sc. and M.Sc. degrees. In particular the strong and increasing demand for computer experts in industry has accelerated this development in Computer Science. However, a key problem has to be resolved at the senior level of undergraduate education. After acquiring programming skills and knowledge about program structures, algorithms and application areas that are immediately needed in professional practice, students are supposed to acquire an overview of research and development in core areas of Computer Science. At the same time they are expected to develop the basis for expert understanding needed in their professional lives. This should enable students to catch up independently with the rapid changes in Computer Science. In the traditional European curricula these objectives are achieved through graduate coursework as there is insufficient time during the period of undergraduate education to cover all the topics. It has been difficult, often unsatisfactory, both for instructors and authors of textbooks, to follow the traditional format of teaching because the presentation, albeit paramount, has frequently to be superficial.

We propose the novel approach of condensing the contents of a subject area and filtering out the essential substance and methods of their treatment as defined by a subject expert. We model teaching and learning strategies according to the characteristics of the essential elements (processes and methods) thus combining the reduction effect with a customised teaching and learning method. This results in a deep, hands-on experience in the subject area. These methods apply to all subject areas in distributed computing. They are equally relevant for revising multimedia and CSCL methodologies in Life-Long Learning programmes as well as in Distance Teaching (Bourguin and Derycke 2001; Koschmann 1996; Restsa et al 1999; Papert 1993; Brereton et al 1998; Gifford and Enyedy 1999). In these areas, most programmes and courses have to adhere to:

- a rigid compaction of the material and its treatment;
- multimedia based CSCL methodologies.

2. CASE STUDY

In a course on operation systems a key subject area is storage management. Processes are regularly stored on hard disks yet have to be resident in main memory for execution. Three major system functions interact to support the execution: *Main memory management, disk management,* and *long-term scheduler* (Silberschatz et al 2000). The first two attempt to manage their space *as efficiently as possible, to accommodate as many processes as possible,* at the shortest possible notice. The latter goal allows for a *high amount of concurrent execution.* The long-term scheduler determines which executable part of a process should next be loaded to main memory for execution. Such executable sub-processes (and necessary data) may be swapped into main memory or back onto disk. This only becomes known at execution time. Since the computation may change the volume of process data, the available space in main memory or the required space on disk, are dynamic parameters for management functions.

The long-term scheduler learns about (dynamic) priorities only when it is scheduling processes on disk to be transferred to main memory. The same is true for the speeds and priorities. In turn, execution times for the three system functions are not predictable - nor are their speeds and space requirements. Thus, although these *system functions* are sequential subprograms under the (centralised) control of a sequential monitor process they *appear to co-operate independently and autonomously.* The co-operation comes about as they execute their algorithms in an interleaving fashion, through the invocation by the monitor. The same is true if one abstracts from the service layer and considers user processes that share resources.

The unpredictability of system or user process behaviour means that:

The essence of storage management is to design the corresponding service functions as autonomous processes that collaborate in a framework where no assumptions on (relative) speeds, time of occurrence, priorities or centralised control could be utilised.

The problems involved are the key problems behind efficient storage management and are very sophisticated. If their nature is fully understood it is easy to perceive both the motivation and the interactional framework of the algorithms involved. If one tries to condense (traditional) storage management to its essence then the diverse algorithms can be dropped as a detail of secondary relevance.

Therefore it is very important to formulate a novel CSCL approach based on matching the subject structure with a novel teaching and learning role behaviour. Concentrating on the essentials of the subject will give the students the ability to deduce the importance of the key concepts, a desirable ability in any research and development environment.

2.1 The Dining Philosophers' Problem

The Dining Philosophers' Problem (Winowski 1981) is a vivid example of this class of problems. Consider five philosophers, who spend their lives thinking and eating, sitting at a round table with a bowl of spaghetti in the middle. The table is laid with five forks, one between each neighbouring philosopher (Figure 1). When a philosopher eats, he needs his left and right hand forks - hence he prevents his neighbours from eating. When the philosophers think they do not eat and therefore do not need their forks.

Figure 1. The Dining Philosophers

This is a simple representation of a finite set of concurrent, autonomous processes (philosophers), which run mainly independently (thinking). Occasionally they perform an action (eating), called critical section, during which they need exclusive access to a subset (their two forks) of the resources (all forks).

The solutions must meet the following conditions:

- *mutual exclusive access to the shared resources* (each fork can be used by only one philosopher at the same time);
- *no assumptions on the speed, the execution-time, and the need for resources* (the philosophers represent autonomous processes);
- *symmetry* (all philosophers have equal rights);
- *fairness* (no starvation);

- *no deadlocks* (the solution must not create a situation in which the philosophers block each other so that one or all of them can move any more towards resource access);
- *efficiency* (minimum overhead and waiting-time).

A simple solution is as follows. A philosopher who wants to eat, takes his left hand fork, if available, otherwise he waits for it. Then he tries to pick up his right hand fork in the same way. This solution fails when all philosophers get hungry simultaneously. Each grabs his left hand fork and will wait for his right hand fork endlessly, because nobody will give back his left hand fork. From this moment on the group is stuck - a situation called deadlock. Deadlock situations may be prevented if the philosophers give back the first fork when the second one is not available, or the philosophers take the first fork only if they are sure that the second fork is available as well.

The students should experience how co-operation must be organised so that all conditions are met. They shape solutions by assuming the role of philosophers. In a classroom environment the instructor mediates the discussion. Only a few students can play the role of the philosophers. The others perform the student role discussing the solutions. A CSCL environment allows all students to interactively perform the roles in this scenario and act like collaborating autonomous processes.

Once a deadlock-free solution has been developed students are encouraged to explore other aspects like starvation-freeness. This CSCL approach that realises interaction among students and instructor is greatly superior to traditional classroom teaching.

3. ROLE-BASED ADAPTIVE LEARNING ENVIRONMENT

Our novel approach is centred on a role-based communication model formulated out of Content-Centred Activity. We assign simple and compound roles to all those involved in the teaching and learning environment. All participants are autonomous processes, there is no central command unit that co-ordinates the syntax of the communication pattern among different actors. While playing a role, a number of bi-directional communication channels are established between the actors on which information, knowledge and experiences are exchanged. These communication channels play an important role in understanding the scenario and the context relating to a concept. We therefore propose a framework that provides an abstract modelling of the syntax of roles and communication channels and then performs content-specific refinements to add the semantics to the syntactic model.

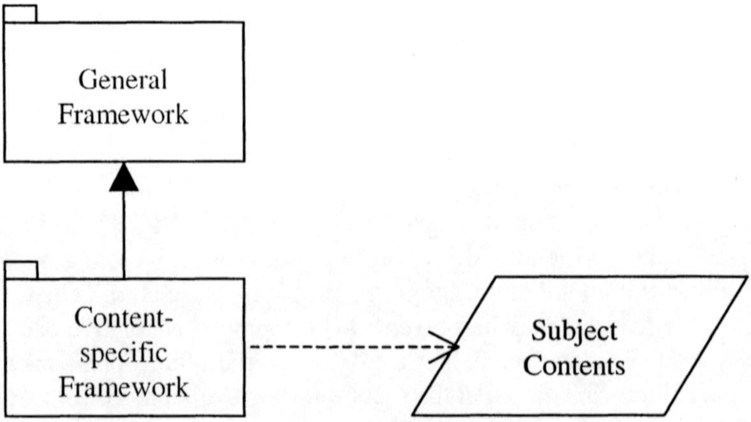

Figure 2 The Role-based framework

The idea of a role-based framework is to model instructional processes in a university lecture environment to minimise the misunderstanding of concepts presented in a lecture. We have two models - the general framework which abstracts the syntax of roles and communication channels in a learning environment, and the content-specific framework which provides the semantics to the roles and communication channels depending on the contents of the subject to be taught (Figure 2).

3.1 General Framework - Roles

The general framework realizes three roles in a learning environment

An Educator is responsible for imparting the concepts related to any subject area. He organises the contents according to a simple principle - "To present the unknown concepts in the form of known concepts".

A Broker is responsible for reducing the burden on, and the amount of queries to, the educator. Students should first contact the broker with a request for information they need. A broker, by virtue of his knowledge and experience, tries to reply to their requests. Should he not have the information in his knowledge base he contacts the educator for help. In a learning environment many brokers can exist.

The Student is at the focus of a learning environment. A large number of students exist in a learning environment and they form learning communities or groups to get a better understanding of the information they receive on a communication channel from the educator and brokers. Communication between students is the most important factor that helps in understanding a concept. The students are asked to assume the roles of different concepts and then simulate the context of the environment to propose solutions to the problem scenario. After performing the roles, they can contact their broker and receive feedback on the simulation.

3.2 General Framework - Communication Channels

The general framework emphasises four communication channels in the learning environment (Figure 3).

Educator-Student: On this communication channel the information, knowledge and experiences of the educator flow to the students.

Broker-Student: The Educator-Student channel is time bounded while the Broker-Student channel is time relaxed. Students can simulate the concepts in a more involved and deeper way. A broker provides feedback to a smaller community of students.

Educator-Broker: A Broker creates a pedagogical script (by discussing with the educator) for the students so that they perform their roles in an effective and informative manner. He provides feedback about the broker-student channel to the educator.

Student-Student: Ideas, algorithms, and discussion flow on these channels as students perform the roles assigned to them by the educator and brokers.

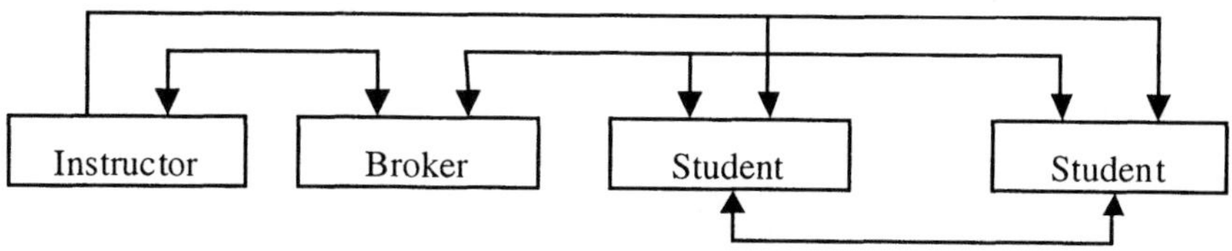

Figure 3. Communication channels in the learning environment

The efficiency of the communication channels is measured by the quantity and quality of information, knowledge and experiences that are exchanged. The quantity depends on the role a participant is playing. However the quality of a communication channel depends on the abilities of the person performing the role, the scenario in which the role is being performed and the number of people involved.

3.3 Content-specific framework

The co-operation among Autonomous Processes, as discussed in the case study, is the essence of teaching operating systems. It is reasonable to consider the students, brokers and educators as autonomous, collaborating processes when teaching operating systems.

Kernel/Educator

In this framework we consider the educator giving a lecture to act as the kernel. The students play the role of user processes. The kernel is the most important process on a computer (Silberschatz et al 2000) since the other processes depend on it for scheduling and execution. The kernel provides the interfaces for the co-operation of user processes with the kernel and among each other. During a lecture the educator plans, co-ordinates and manages the flow of presentation and discussions and allows questions.

Virtual Machine/Broker

Tutorials offer another classroom environment. Brokers present solutions to homework assignments and lead discussions on unsolved problems. We map the brokers to kernels on virtual machines. Virtual machines create an illusion that a process has its own processor and kernel. They enable processes to perform actions without directly interacting with the real kernel.

Scheduling/Discussion

In modern computers the CPU executes multiple processes concurrently by switching among them. CPU scheduling deals with the problem of deciding which process is to be run. Mapping the CPU-scheduling to chairing a discussion, where only one person may speak at a time, illustrates a problem of the subject area.

Discussions in group-learning environments are a representation of process synchronisation and resource scheduling problems in operating systems.

Context Switch/Questions

Switching the CPU to another process requires a context switch. The kernel saves the state of the old process and loads the saved state of the scheduled process. Although necessary, context-switch time is pure overhead - the system does no useful work while switching.

A comparable situation is a student asking a question during a lecture. It takes time for others to become aware of the student's inner communication context, which is necessary to understand the question entirely. Modern operating systems solve this problem by introducing threads. Processes can contain multiple threads of control, which share most of their information. Communication and context-switches between threads of the same process have less overhead due to the mutual context. We consider the working groups as processes and the students as threads.

Threads/Students

The students play the role of collaborating threads. In a classroom environment with a large number of students the communication overhead can exceed a tolerable limit. The students form teams and act like the threads of a process. They are familiar with each other's knowledge obtained by learning together. If the group wants to take part in a discussion the speaker addresses the other students or the educator. This concept makes an efficient discussion with a large number of students possible. The educator has only to pay attention to the speakers.

Communication

Collaborating processes can communicate using shared-memory or message passing. Communication by shared-memory means one process writes data into a shared buffer, another process reads from the buffer. Students perform similar actions when they collaborate on homework assignments. They read from and write on a piece of paper. Message passing is a form of inter-process communication very similar to human conversation. Processes send messages to each other. In both cases the students learn about the mechanism that allows processes to communicate and synchronise their actions by acting like collaborating processes.

4. CONCLUSION

We have outlined a novel principle of teaching and learning under heavy time or communication constraints such as in Distance Teaching or Life-Long Learning programmes. Our approach is based on matching the structure and behaviour in the teaching and learning environment with the structure and behaviour of the subject matter processes. We have explained in our example of an undergraduate-level course on operating systems how we distill the true nature of the problems and processes out of the traditional teaching material. This condenses the presentation and provides an adequate, hands-on experience and understanding of the essence of operating systems. Well-known CSCL methods have been combined with a novel role-based model of the teaching and learning environment. We have improved the instructional productivity and the learning quality for both students and instructors.

We are working on a completely interactive teaching and learning environment based on individual laptops with wireless connections. This will be the subject of forthcoming publications.

REFERENCES

Bourguin, G. and Derycke, A. (2001) Integrating the CSCL Activities into Virtual Campuses: Foundations on a new Infrastructure for Distributed Collective Activities. In *Proceeding of Euro-CSCL 2001*.

Brereton, P., Lees, S., Gumbley, M., Boldyreff, C., Drummond, S., Layzell, P., Macaulay, L. and Young, R. (1998) Distributed group working in software engineering education. In *Information and Software Technology*, No. 40. pp. 221-227.

Gifford, B. R. and Enyedy, N. D. (1999) Activity Centered Design: Towards a Theoretical Framework for CSCL. In *Proceedings of Computer Support for Collaborative Learning*. pp. 189-197.

Koschmann, T. (1996) Paradigm Shifts and Instructional Technology: An Introduction. In *CSCL: Theory and Practice of an Emerging Paradigm*, T. Koschmann (ed.), Lowrance Erlbaum Associates.

Papert, S. (1993) *The Children's Machine: Rethinking School in the Age of the Computer*. Basic Books, New York.

Resta, P., Christal, M., Ferneding, K. and Kennedy, A. (1999) CSCL as a Catalyst for Changing Teacher Practice. In *Proceedings of Computer Support for Collaborative Learning*. pp. 488-495.

Silberschatz, A., Galvin, P. and Gagne, G. (2000) *Applied Operating Systems*. John Wiley & Sons, Inc., 1st edition.

Winkowski, J. (1981) Protocols of accessing overlapping sets of resources. In the *Information Processing Letters*, Vol. 12, No. 5. pp. 239-243.

KOLUMBUS: Context-oriented communication support in a collaborative learning environment

Thomas Herrmann and Andrea Kienle
Informatics & Society, University of Dortmund, D-44221, Dortmund, Germany

{thomas.herrmann; andrea.kienle}@udo.edu

Key words: Collaborative Learning, Communication, Learning Environments

Abstract: Computer supported collaborative learning occurs primarily by means of communication. Successful computer supported co-operation is only possible if communication among learners and between learners and teachers is appropriately supported. We describe communication on the basis of a context-oriented communication model that focuses on the dialogical communication and mediation of context. The model leads to an abstract identification of general communication tasks amongst collaborating learners and between them and their teachers. The resulting list of tasks can be generally used to derive requirements for the support of communication in collaborative learning environments. This is demonstrated with the features of the prototype system "KOLUMBUS", which is used for experimental investigation in university courses. It realises the support of all identified communication tasks. The paper ends with a brief summary of results taken from experiences at a university seminar as part of a larger research work.

1. INTRODUCTION

An increasing number of developments can be assigned to the paradigm of computer supported collaborative learning (CSCL). The most important concept of collaborative learning is the active role of the learner and Koschmann (1996) and Wiburg (1995) recognised a paradigm shift in this development towards a self-responsible learner.

For CSCL-Systems, the support of interaction is mostly focused on communication. Rather than directly experiencing a situation or learning by observing learners interact at different locations and during different time slots. Appropriate methods which support communication among learners and between learners and teachers are decisive factors in the success of collaborative learning environments. In the past the area of support for communication was insufficiently investigated.

To improve CSCL environments we focus on those types of communication theory where communication is understood to be symbolic interaction between identifiable persons who can be directly addressed in the course of dialogues. The mass distribution of instructions (as part of an expression) is not included in this perspective if there is no opportunity for learners to realise their success in learning and achieve mutual understanding through the process of social interaction.

We have developed a prototype system called KOLUMBUS which is strictly oriented towards the support of interactive, mainly asynchronous communication and the conveying of the context which completes the communication processes. This idea of strict orientation to communication support was inspired by the experience with other CSCL systems. *BSCW Shared Workspace System* (Appelt and Mambrey 1999) allows easy upload of documents, however, the presentation of content takes place in the system in which they were developed e.g. Microsoft Word. Furthermore, communication is not sufficiently supported in this system as it occurs outside the workspace. *CSILE* (Computer Supported Intentional Learning Environments), on the other hand, allows all participants to add any desired information to the system and this content can be used as easily accessible context information. A disadvantage is that participants have to use html commands to add their content to the system if they wish to produce readable, pleasant presentations that foster understanding (Scardamalia and Bereiter 1996). *Gentle*, allows readable versions of content to be produced which can then be annotated but it is neither possible to limit the group of recipients nor to extend the set of responsible authors (Dietinger and Maurer 1998). In *WebGuide* (Stahl and Herrmann 1999), statements or annotations can be added to the content on the system. However, these contributions did not relate closely and there was a lack of convergence in discussions. We concluded that structured negotiation and the facilitation of discussions should be supported in CSCL systems.

KOLUMBUS is based on these results and is designed to improve communication-based learning in universities. The educational goal is twofold. Firstly, students should benefit from the experience of others by conducting research on a specific topic (e.g. the conditions under which the New World could be discovered by Columbus) and knowledge of these topics should be deepened by discourses between students. Secondly, students should learn how social learning can be conducted through the medium of computer networks under asynchronous communication conditions. This requires more explicit interaction and documentation as well as a high degree of discipline.

2. CONTEXT-ORIENTED MODEL OF COMMUNICATION

In computer science, models of communication are generally characterised by the sender-recipient model (Shannon and Weaver 1949). In this model the sender's message is transmitted to the recipient using a channel. The message is encoded before and decoded after the transmission. In psychologically-oriented research, however, the influence of both communication partners on the communication act and its contribution to increase shared understanding (Clark and Brennan 1991) and the relevance of context for the communication (Ungeheuer 1982) are more important. This leads to a kind of communication model which is focused on how mutual understanding can succeed in dialogues. Its origins are rooted in neo-constructivism (von Glaserfeld 1996) and biological epistemology (Maturana and Varela 1987).

Maturana and Varela in particular point out that the models where something such as information is transported from a sender to a receiver are completely inappropriate. Sender and receiver are closed systems with respect to information processing. A communicator selects from the universe of his/her beliefs what he/she wants to say and the recipient selects, with respect to his/her universe of beliefs what he/she wants to understand. These universes of beliefs may be considered a kind of inner context and this is closely related to the outer context they can share in their perception. One important part of the context is the reciprocal knowledge of the other. Our communication model is based on the assumption that only the availability of context can give the communicators a feeling of whether their communicative acts are successful not.

2.1 Activities of the communication partners

A communicator selects the idea which has to be represented (**what**) and selects the appropriate paraphrase. This means deciding **how** the idea has to be presented to specific recipients to be easily understood. Only then can an expression be uttered. The recipient creates a mental representation of the expression and transforms this representation into an impression related to the recipient's universe of beliefs. Producing an impression can only succeed if the recipient inter-relates his/her pre-knowledge with the representation of the expression. It is vital for learning that the impression is integrated into and made a relatively permanent part of this universe. Some articles deal with the fact that learners convert information into knowledge on the basis of their already existing knowledge (Feltovitch et al 1996). It is the task of both communicator and recipient to check whether the communication act was successful or has still to be completed.

Context has an important relevance for several activities of the communicator and the recipient. Remember that context combines many different aspects - the physical situation, the cultural and geographic constellation, the characteristics – especially the knowledge – of alter ego, the expressions already uttered by the participants in communication and the extra-communicative behaviour of the participants. This extra-communicative behaviour is especially important because it is independent from symbolic meanings.

The special relevance of context is seen in certain activities of a communication act.

- (Selecting what has to be communicated) By being aware of the context and by having a model of the knowledge a recipient already has, or can easily derive from the context, the communicator can decide whether it is sensible to communicate an idea for purposes such as teaching, giving an orientation or entertainment.
- (Selecting the appropriate paraphrase) The context gives hints about how explicit an expression should be. The communicator can decide whether a certain aspect of his idea has to be explicitly expressed or is evident because of the context which is shared with the recipient. Furthermore, the communicator uses the context to decide which level of language might be understood or what metaphors or comparisons are appropriate.
- (Making an impression) By relating the representation of the expression to the context, especially to the assumed knowledge of the communicator, it is easier for the recipient to find out what the communicator might have meant. Context completes the representation of the expression. Furthermore, the recipient can identify whether he/she needs additional information from the communicator or not.
- (Learning) The context can be employed to make implicit decisions on whether an impression is sufficiently relevant to be relatively permanently integrated into the universe of the recipient's beliefs.
- (Avoidance of misunderstandings) Observing the context – particularly the ongoing communication but even more importantly, the extra-communicative behaviour - develops awareness of misunderstandings.

3. REQUIREMENTS FOR THE SUPPORT OF COMPUTER MEDIATED COMMUNICATION

The mutual context can be reduced during online communication compared with face-to-face communication. Only certain parts of the context of the communication partners can be mutual (knowledge of physical laws, geographic constellations). Consequently, communicators should ensure the relevant context is available for the recipient. We can differentiate between the following situations with respect to collaborative learning:

1. A teacher or a student collects material (content) for their own work. While doing so, they do not connect any communicative intentions with this material. When the teacher distributes a task description to the students, or when the student makes a proposal to his fellows, these are communicative acts. The material serves as context for these acts and is provided by the communicators themselves. They do not have to feel responsible for the communicative quality of this material.
2. Teachers or students formulate an expression and realise that their potential recipients might not have enough context at their disposal to understand what is meant. Therefore, the communicators make additional material accessible which creates the relevant context - an e-mail attachment or the inclusion of a URL. Another solution might be to insert a communicatively meant expression directly as an annotation into the material.
3. In some cases, the participants of the collaborative environment may realise that it will be difficult for the recipients of a message to combine it with the available material or that the availability of context is insufficient. They may decide not to rely on the recipients' context perception but to express the relevant parts of the context in their own words and integrate it directly into the expression. Thus, they make context explicit.

The main idea of supporting the activities of communication partners in computer mediated learning processes involves the individual and flexible presentation of content.

To note the knowledge of the recipient, the communicator can inspect, for example, a web-based collection of the recipient's information. Using links and by exploiting the free formulation of contributions, the communicator can explicitly describe the relevant context. Since it is hard to anticipate the amount of context required, it is sensible to offer possibilities for splitting the conveyed context into several parts that can be selected flexibly by the recipient. Hide and show mechanisms are excellent for concealing unimportant aspects. Table 1 summarises the activities of the communicator and the software components of KOLUMBUS.

Table 1. Support for the activities of the communicator

Activities of communicators	Software components in KOLUMBUS
Being aware of the knowledge of the recipient	Offering **perspective** where information can be browsed related to the recipient and available in the system. Support of browsing this information by meta-information, search, rating and filter mechanisms and awareness features.
Making context available	Offering possibility to **annotate** communicative contributions directly to those parts of content to which they are related. Adding links to messages which refer to context. Offering means to support segmentation of content into small items.
Omitting known facts and concealing unimportant aspects	Support **divisibility of content** by hide and show mechanisms based on means for segmentation and tree views.
Compensating for the reduced possibilities of expressions	**Multimedia** based possibilities of representing different layouts of content. **Different communication channels** e.g. chat.

The recipient can use his own collection of information to relate the expression to his/her own knowledge and the collection of the communicator to note his knowledge. All the components mentioned in Table 1 are important for the recipient.

Awareness features support the identification of the mediated context. To reconstruct the context, easy access to documents is needed. To support requests for further information support for interactivity has to be offered.

4. THE COLLABORATIVE LEARNING ENVIRONMENT KOLUMBUS

The aim of KOLUMBUS is to provide optimal integration of context presentation and support of communication by offering all the above components. Figure1 shows a process-oriented view of the KOLUMBUS features and their interrelation.

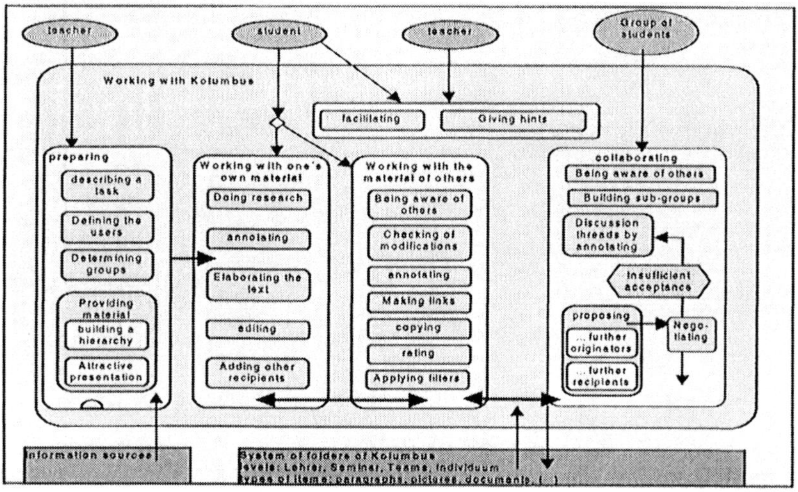

Figure 1. Usage process using the modelling method SeeMe

4.1 Preparing and working with one's own material

A lecturer prepares a seminar by describing a task, determining who has to participate and identifying how these participants may be grouped. KOLUMBUS facilitates the flexible presentation of content, offers two kinds of views of the content and permits its easy upload.

Students build up and improve their own material by collecting, annotating and editing. This can be enhanced by the exchange of information fostering collaborative working.

The central feature of KOLUMBUS is the representation of content in **small units** which enables students to use the stored content very flexibly. The smallest unit, or item, can be a text, picture, binary number, link or annotation. The content can be presented as a hierarchical structure of items viewable in a web-browser. Items can be inserted at any level of the hierarchy and the teacher, as well as the students, can build a hierarchy of their material. All existing functions (annotate, rate, copy, shift, change) can be applied to every item.

The system provides two different views of content. In **treeview** each item is represented as a node in a hierarchical tree-structure (Figure 2) whereas **paperview** illustrates the content in a more visually attractive and readable way. Different types of presentations can be combined to form a single document. In paperview it is possible to expand or omit content of every item, and all functions are available. Uploading content is easy - no HTML knowledge is necessary. The menu providing access to the system's functions is bound to items so further items may be added at the place from which the participant called the menu.

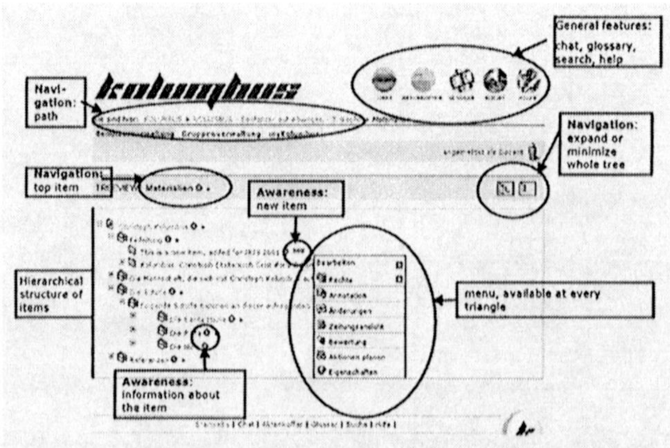

Figure 2. KOLUMBUS - a treeview

As it is necessary to disperse longer content into many items to enable the flexible presentation and use of the content, we have developed two main ways of adding content. On the basis of a DTD, an XML-document can be dispersed into single items and using a predefined Microsoft Word template an XML-document (whose titles, sections or tables are recognized as KOLUMBUS items) may be created.

4.2 Working with the material of others

Students can identify and check new or modified items and be aware of other's activities by using the awareness features (meta-information, search, rating and filter). In the treeview, new and modified items can be recognized.

Search functionality is provided. A communicator might use this to identify the knowledge of the recipient and a recipient might use it when including knowledge or reconstructing the context. Searching for "annotation" retrieves all contributions to discussions. A rating function allows users to give feedback and a filter restricts the visible items to the chosen group of authors and/or recipients.

Using the annotation function participants give feedback on others' materials. The concept of items enables them to annotate directly in the content and makes the context explicit. Discussion threads are developed by annotating annotations. KOLUMBUS also offers synchronous communication through the chat function - extremely valuable for requesting further information.

To support the collaborative process of learning, a concept of different **perspectives** is realized. In KOLUMBUS, a perspective is defined by three characteristics: its content, the set of originators responsible for the content and the set of recipients. Extending the group of recipients allows an originator to publish his content to others.

4.3 Collaborating

Group results are developed by the process of taking responsibility for certain items. The originator initiates a negotiation by proposing further originators. The proposed originators vote for or against their originatorship for the item in question and if a certain percentage agrees with the proposal, the group of originators will be extended. Thus the group achieves convergence on a certain set of items. If a group of students is the originator of an item and a single person wants to extend the group of recipients, the group members also have to negotiate the proposition of further recipients. Every participant can be aware of the progress of the collaborative learning process; a request for a negotiation for example shows a concluding step in the process.

5. CONCLUSION: FIRST EXPERIENCES

Since April 2001 we have tested KOLUMBUS in a seminar environment with 16 students. Its use was evaluated through interviews and the analysis of logfiles.

The possibility of observing others' material and their usage of the system was considered extremely helpful in fulfilling personal tasks. The concept of annotations was highly accepted to support the communication during the feedback process. Students added annotations to singles items and confirmed they used existing content as context information that did not have to be explicated again. As recipients of annotations they indicated understanding was improved by reading the content and the annotation at the same time.

Some students described problems finding the annotations. Awareness features like "new" tag were not sufficient. Students suggested a list of new items should be displayed on the top screen immediately after login or an e-mail notification. They also requested a feature for displaying all annotations to their material in a single view. Further research is necessary to find out how features of KOLUMBUS, which seem to be too general, should be specified to reflect the underlying scenario.

Despite these shortcomings, our evaluation concluded that the integration of context presentation and support of communication offered by KOLUMBUS improved communication-based learning.

REFERENCES

Appelt, W. and Mambrey, P. (1999) Experiences with the BSCW Shared Workspace System as the Backbone of a Virtual Learning Environment for Students. In *Proceedings of ED Media '99*, Charlottesville. pp. 1710-1715.

Clark, H. H. and Brennan, S. E. (1991) Grounding in Communication. In Resnick, L. B., Levine, J. M and Teasley, S. D., *Perspectives on Socially Shared Cognition*. American Psychological Association, Washington, DC. pp. 127-149.

Dietinger, T. and Maurer, H. (1998) GENTLE - General Network Training and Learning Environment. In *Proceedings of the Ed-Media 1998*, Freiburg.

Feltovich, P. J., Spiro, R. J., Coulson, R. L. and Feltovich, J. (1996) Collaborative Within and Among Minds: Mastering Complexity, Individually and in Groups. In *CSCL: Theory and Practice*, T. Koschmann, (ed.), Lawrence Erlbaum Associates, New Jersey. pp. 25-44.

Glasersfeld, von (1996) *Radikaler Konstruktivismus*. Suhrkamp, Frankfurt am Main.

Koschmann, T. (ed.) (1996) *CSCL: Theory and Practice*, Lawrence Erlbaum Associates. New Jersey.

Maturana, H. and Varela, F. (1987) *Der Baum der Erkenntnis*. Scherz. Bern, München, Wien.

Scardamalia, M. and Bereiter, C. (1996) Student Communities for the Advancement of Knowledge. In *Communications of the ACM*, Vol. 39, No. 1. pp. 36-37.

Shannon, C. E. and Weaver, W. (1949) *The Mathematical Theory of Communication*, The University of Illinois, Urbana (Illinois).

Stahl, G. and Herrmann, T. (1999) Intertwining Perspectives and Negotiation. In *Proceedings of Group'99 International conference on Supporting Group Work*. Phoenix, Arizona. pp. 316-325.

Ungeheuer, G. (1982) Vor-Urteile über Sprechen, Mitteilen, Verstehen. In *Kommunkationstheoretische Schriften* 1, G. Ungeheuer (ed.), Rader, Aachen. pp. 229-338.

Wiburg, K. M. (1995) An Historical Perpective on Instructional Design: Is it Time to Exchange Skinner´s Teaching machine for Dewey´s Toolbox? In *Proceedings of the First International Conference on Computer Support for Collaborative Learning (CSCL ´95)*, J. L. Schnase, and E. L. Cunnius (eds.). pp. 385-391.

Teaching Social Informatics as a knowledge project

Iver Jackewitz, Michael Janneck, Detlev Krause, Bernd Pape and Monique Strauss
Department for Informatics, University of Hamburg, Vogt-Kölln-Straße 30, D-22527 Hamburg, Germany

{jackewitz|janneck|krause|pape|strauss}@informatik.uni-hamburg.de

Key words: Collaborative Learning, Curriculum Policies, Didactics of Informatics, Higher Education

Abstract: Teaching Social Informatics poses the challenge of portraying future scenarios of technology development and use as well as criticising them. We envision meeting this challenge by offering students an authentic, interdisciplinary, educational framework, providing real-life, hands-on experiences. In this paper, we will first develop the idea of "knowledge projects" as a means to encourage such educational settings. Then we will focus on our efforts to develop a Web-based software system, named CommSy, designed to support communication and co-ordination in learning communities involved in knowledge projects.

1. INTRODUCTION

The term "Social Informatics", coined by Kling (1999) was defined as the interdisciplinary study of the design, uses and consequences of information technologies that takes into account their interaction with institutional and cultural contexts. Enabling students to better understand the interaction of information technologies and their fields of application is a central issue in Social Informatics curricula, because, in this subject area, we train future specialists who will play an active role in developing new information technology solutions for people in other fields. Teaching Social Informatics has to deal with the ambiguity of promising future scenarios of technology development and use on the one hand and with harsh critics of these scenarios on the other.

One promising scenario, for example, is the 4A scenario:

"Any one can get any document anytime and anywhere. There are instantiations of 4A such as providing any researcher with all of the documentary materials that they want for their research, even if they are traveling for a month; or providing any doctor with a complete medical record for any patient, anytime, anywhere." (Kling and Star 1998)

Living and working in the 4A scenario would require various professional, social and technical skills in dealing with unfamiliar problems and becoming acquainted with ever changing circumstances. This comprises, for example, the ability to get to know colleagues, grasp their perspectives on current problems, and to take up work done by them, all in a short period of time. Therefore, it also seems essential that we should make effective and efficient use of software tools that support such project teams. However, scenarios like 4A are criticized because they too easily trivialise the interaction of design, uses and consequences of information technologies with their context:

"by homogenizing people and places into 'everyman' and 'everywhere'. The various roles that people play in work groups are ignored and stereotyped. The ways that organizations structure information is also treated only as a barrier, unless materials are accessible 4A. The different kinds of resources (and skill sets) of organizations and groups are also all homogenized in 4A scenarios." (Kling and Star 1998)

To offer our students the opportunity to learn the necessary professional, social and technical skills to live and work in a probable 4A scenario, and to take a critical stance on the prospective development at the same time, we consider it is not sufficient to merely present this ambiguity. In our opinion, students should have the opportunity to gain experience in technology use, in new forms of work organization and in related ways of organizing their lives during their studies. To offer our students authentic educational settings, we regard studies in Social Informatics as a "knowledge project". In this paper, we will first describe what we mean by "studying Social Informatics as a knowledge project". We examine two views on knowledge projects that guided us in our efforts to develop software support for teaching Social Informatics and then focus on the resulting software system *CommSy*.

2. KNOWLEDGE PROJECTS

What is a "knowledge project"? Let us start defining the term by explaining its constituent parts. We consider "knowledge" to be a context-dependent representation of information (Schultze 1998) and "projects" we define as learning opportunities that emphasize practical, problem-centered, activity-oriented learning (Gudjons 1997). Therefore, when we speak of "knowledge projects", we are basically talking about context-based knowledge obtained in hands-on, real-world settings, facing real problems which are of actual importance to students and to other people.

We identify four major aspects of knowledge projects:

Acquiring knowledge. As mentioned above, we do not consider knowledge as context-independent. However it is not completely individualized either. Knowledge may be established by communities and is carried on through traditions and customs. Imparting knowledge is thus not an activity that is passively received by one or more learners. Acquiring knowledge has to be seen as an active endeavour integrating "new" pieces of information into existing structures of knowledge and experience in which social and group experiences play a major role. From this viewpoint, faculty staff are seen as facilitators (Rogers 69) who try to spark off and encourage individual and group learning processes - rather than "classical" lecturers who strive for a close match of student know-how with their own.

Multiple players and multiple perspectives. Learning and acquiring knowledge in a university context always involves multiple players - especially students, teachers and professionals from the respective vocational field - with different views and experiences. These multiple perspectives should not be ignored or streamlined. They should be taken into account and used to enrich the learning experiences of those involved. Thus, studying informatics as a knowledge project means that different standpoints and experiences are emphasized, dealt with explicitly and tied together into a broader perspective. This implies on the one hand that the students' viewpoints and experiences are taken as seriously as those of "experts", but on the other hand it means that different persons never learn - or should be forced to learn - exactly the same things. Individual experiences serve as a starting point, an anchor from which further learning experiences are developed in the respective context and can be shared with others.

Crossing boundaries. While it is important to have multiple perspectives within one community of practice, we also emphasize the importance of multidisciplinary exchange and communication. In modern vocational fields professional barriers have become permeable. A lot of tasks require a broad range of different skills from different professional backgrounds. We emphasize the importance of gaining different vocational competencies not only in professional life, but also during university education.

IT support. IT support in university education should "support" those involved in a way that actually makes work more efficient and effective or allows for new ways of communication, cooperation or storage of information. Another important aspect of IT support in university education lies in the opportunity for students to gain competencies in the use and handling of "new" media such as the Internet.

The concept of university education as a knowledge project implies two perspectives. Firstly, it can serve as a model for organizing singular learning activities (lectures, study groups). This emphasis on project-based learning lets students work on complex problems in small teams over an extended period of time (Pape et al 2002). Secondly, we pursue the goal of understanding and organizing university courses of study - in our case particularly Social Informatics - as a whole as knowledge projects. This means we have to bridge the gaps between separate courses and strengthen connections between different parts and contents of university study courses. The keywords of continuity, transparency and coherency regarding different content and methodology in university education are vital in this respect.

3. SOFTWARE SUPPORT FOR KNOWLEDGE PROJECTS

In our research and software development process, we strive to provide software support for knowledge projects. Since there is always a set of different traditional and "new" media used in educational settings, we don't try to replace them with a system which claims to cover all communication needs in university teaching. We consider this neither feasible nor desirable. In acquiring competences in the use of technical support for communication and coordination, it is more important to strive to develop the capability of handling a variety of technologies. Knowing which medium is appropriate to which communicative need and knowing how to express this need in the selected medium are necessary skills for living and working in a networked world. That is why we concentrate on the two perspectives on knowledge projects as learning environments in our research and development.

By concentrating on knowledge projects, we emphasize the situated nature of learning processes. Learning processes do not happen anytime but at a particular time, consisting not just of anyone but of particular persons, motivated not by any reason but by particular reasons and not working with any documents but with particular ones. Our software development aims to support faculty and students to take part in knowledge projects in an active and reflective manner.

4. COMMSY

CommSy (CommSy 2001) stands for Community System and is a web-based system to support communication and coordination in learning communities. In its latest version *CommSy* comprises three components: *CommSy Project Rooms* support close cooperation in closed work groups with 10 - 30 members. *CommSy Archive* provides a collection of study materials and student work which is partly accessible worldwide. *CommSy Campus* links the archive and the project rooms, offers information on courses and extracurricular activities and thus provides central access to the system.

CommSy Project Rooms are designed to support singular learning activities as knowledge projects (Pape et al 2002). By virtue of its functional scope, the system supports central activities of a learning project as outlined by Gudjons. It supports communication in multiple ways: News and events can be announced, specific topics can be discussed in discussion forums, each member has a personal homepage to present him/herself to the group, and annotations can be made to every item in a *CommSy Project Room.* Working material can be collected in a simple reference manager and put in context by attaching them to any other item (such as an announced event). A group-editor is available for cooperative writing of HTML documents.

The design principles of *CommSy Project Rooms* demand (and allow) that the members of a community establish a shared understanding of their working context. For example, they can attach intermediate results to any item in the system, but they must also find ways of making each other aware these results exist. Members have to relate their own communicative purpose to those of others and cannot rely on an automatic facility provided by the software system. In addition, a community must share an understanding about how often the participants use the system and thus read others' items - once a week, every second day or more than once a day. If a communication need appears to be too urgent to fit the expected frequency with which other participants use the system, additional communication media, like telephone or e-mail, have to be used.

Three design principles distinguish *CommSy Project Rooms* from other software products:

- Easy individual use: Enabling individuals to use *CommSy Project Rooms* easily is a prerequisite for any project member to actively engage in the work without having to overcome technical barriers. We achieve this by implementing only required functionality and a simple, recurring structure across the whole system.
- Responsibility in cooperative usage: *CommSy Project Rooms* give special emphasis to user communities rather than individuals: *CommSy*

Project Rooms are exclusively accessible to members of a certain group and each user's name is recorded with every item s/he creates. Every project room can be customized to help build a group identity by colour selection, by choosing a name for that specific *CommSy Project Room*, and/ or by selecting a subset of available functions.

- No concept of roles: "Ownership" is the sole access right in *CommSy Project Rooms*. Only the owner of an item may modify or delete it. Apart from this rule, every member of a *CommSy* is allowed to do everything and to see everything. There is no distinction between students and teachers. Users can be sure that they get the same presentation as everyone else. In *CommSy Project Rooms* we promote good social behaviour in our day to day interaction with students - stressing self-responsibility and commitment.

CommSy Archive is designed to support teaching and learning individuals over a longer period of time. Users can store and find study materials of any kind. It is possible to select material for special purposes, to leave comments on every item of material or to rearrange it. Every item of material can be labeled as relevant for one or more courses and can be categorized "topic", "kind of material", "author" and "user". Additionally, there is a bibliography for any literature that is not part of *CommSy Archive* itself, a catalogue of contacts (institutions, organizations and people) and a glossary.

CommSy Archive is characterised by multiple forms of access for users, who have to deal with rapidly changing topics and perspectives. Most of the archive is accessible only to participants of courses using the system. Selected contents are accessible worldwide but, in order to avoid misuse, only a few authorized people have the right to publish files for worldwide access. As a result, the narrow context for tutorial purposes and the wide context for establishing a broad discourse on Social Informatics are effectively separated. The worldwide access is only a part of *CommSy Archive* situated in a particular learning community, i.e. in a particular knowledge project.

In order to set up *CommSy Archive* as a part of a knowledge project, it is important to integrate its main archiving functions into virtual and real cooperation settings. *CommSy Archive* not only supports information retrieval but also serves as a publishing medium. Student authorship is one of the core principles *CommSy Archive* is designed to support. Students and faculty staff may produce results relevant for a broader discourse and structure material to fulfill different, but specific needs, thus contributing to the context of the future usage of the archive.

Learning communities may take interesting material from *CommSy Archive* into their *CommSy Project Rooms* in order to discuss it more intensively and improve it. Thus, material on *CommSy Archive* potentially meets the criterion of "state-of-the-art" because new versions or new papers from *CommSy Project Rooms* contribute to the archive.

CommSy Campus serves as a portal, providing central access to *CommSy Project Rooms* and *CommSy Archive.* User support and other administrative functions are bundled here. Users - students as well as teachers - may register offered courses and extra-curricular activities here, provide information and study materials relevant for their courses and apply for a *CommSy Project Room.*

A central facility on *CommSy Campus* is a listing of courses and extra-curricular activities in current and earlier semesters, complete with appropriate project rooms and archived study material. Several courses belonging to a common context may be arranged in so-called "mini-curricula". *CommSy Campus* also offers "topics" - a collection and description of relevant subjects which can be attached to courses and archived materials. In addition it offers a notice-board on which announcements can be made.

We consider *CommSy* as a medium designed to support learning communities. While *CommSy Project Rooms* focus on learning communities working together on a particular knowledge project, *CommSy Archive* and *CommSy Campus* strive to give substantial support over a longer period of time and strengthen connections between different parts and contents of study courses. They therefore support university education in its entirety as a knowledge project.

5. FUTURE WORK

Establishing knowledge projects is not only based on the development and use of software systems. In our research and development work, we also consider the following tasks:

- *Embedding the software systems pedagogically.* Since the use of new media in educational settings is seldom successful without changes in the way we teach we intend to concentrate on ways of embedding *CommSy* pedagogically in our future work. We will educate faculty staff in the use of the software system.
- *Establishing knowledge projects as organizational development.* In order to master the technological and organizational aspects of establishing knowledge projects in a university we foster a process of organizational development.
- New didactical paradigms, including devices designed to support learning communities, need careful *evaluation.* Therefore we are monitoring the processes in our knowledge project from different

perspectives (evaluation by learners, teachers, consultants, experts).

Acknowledgements

Christiane Floyd and Horst Oberquelle were extremely generous with their time and inspiration. Thomas Bogner provided very helpful and profound feedback to this piece of work. Special thanks go to the huge number of students who helped us in developing CommSy. Funds for conducting some of the research reported here came from the Zukunftsinvestitionsprogramm of the German Federal Government.

REFERENCES

CommSy Website (2001) [http://www.commsy.org, 30.10.2001]

Gudjons, H. (1997) *Didaktik zum Anfassen: Lehrer/in-Persönlichkeit und lebendiger Unterricht.* Klinkhardt, Bad Heilbrunn.

Kling, R. (1999) What is Social Informatics and Why Does it Matter? *D-Lib Magazine*, 5(1).

Kling, R. and Star, L. (1998) Human Centered Systems in the Perspective of Organizational and Social Informatics. *Computers and Society*, 28(1).

Pape, B., Bleek, W. -G., Jackewitz, I. and Janneck, M. (2002) Software requirements for project-based learning: *CommSy* as an exemplary solution. *Proceedings of the 35th Hawaii International Conference on System Science.*

Rogers, C. (1969) *Freedom to learn.* Merrill, Columbus, Ohio.

Schultze, U. (1998) Investigating the Contradictions in Knowledge Management. *Proceedings of IFIP Working Groups 8.2 and 8.6 Joint Working Conference on Information Systems: Current Issues and Future Changes.*

Using a lecturer's personal Web site to enhance the social interchange among students in an academic course

David Passig
School of Education, Bar-Ilan University, 52900 Ramat-Gan, Israel.

passig@mail.biu.ac.il

Key words: Collaborative Learning, Higher Education, ICT, Social Issues, Virtual University

Abstract: Developments in web-based communication technology have opened up new ways for students at a distance to communicate with their teachers and with each other. The literature covers three types of web-based interactions: learner-content interaction, learner-instructor interaction, and learner-learner interaction. However, the purpose of this study was to examine a newer aspect in these relationships. The aim was to investigate the social impact of a lecturer's personal web site on the personal interchange with his students. The participants were sixty-three students undertaking one of the author's graduate courses at Bar-Ilan University. They were given learning assignments, which required social interaction to complete, using a variety of Peer-to-Peer technologies. The results show that most of the students found the lecturer's personal web site an efficient P2P (Peer-to-Peer) tool that enhanced social interchange outside the walls of the class. Over 85% of the students strongly recommended that personal aspects delivered only on a personal web site should be added into academic courses.

1. INTRODUCTION

Developments in electronic communication have opened up new ways for students to communicate with their teacher and with each other at a distance. There are three main types of interactions detailed in the literature (Moore 1989; Tribe 1994):

- *Learner-content interaction.* This process of intellectually interacting with content results in changes in the learner's understanding, that is, learning.
- *Learner-instructor interaction.* The instructor can stimulate or maintain the student's interest, organise the learning experience, present information, assess and provide information on progress, and provide counsel, support and encouragement. On-line tutoring ranges from comments on the return of an assignment, through posing questions to guide reading and research, to leading discussions on a topic.
- *Learner-learner interaction.* Interaction between one learner and other learners is the way students consolidate their ideas, test hypotheses, and confirm opinions and attitudes. Research into peer interactive learning demonstrates that students who work cooperatively achieve superior results to those working in individualistic or competitive ways. In particular, the practice of students giving and receiving advice and assistance from one another shows a positive relationship with achievement.

To achieve these kinds of learning a variety of learning strategies can be employed. Kibby (1999) lists a number of these, including:

- *Panel discussion* - a group of 'experts' initiate a discussion on a set topic, students join in at any point, and the experts play a diminishing role as the discussion progresses;
- *Brainstorming* - setting forth of ideas without evaluation;
- *Problem solving exercise* - small groups work out a solution to a problem;
- *Teamwork games* - adapted from small group 'bonding' exercises;
- *Role playing* - such as MUDs and MOOs that are forums based on role playing;
- *Simulation* - MOOs can also provide settings for simulations;
- *Co-tutoring* - students are assigned topics which they then 'teach' to another student;
- *Case analysis* - independent study of a case or representative example followed by group analysis;
- *Collaborative writing* - group writing, or individual writing followed by mutual critiquing;
- *Expert in electronic 'residence'* - an expert is available to make presentations, be interviewed and lead discussions;
- *Interview* - questions can be asked of the instructor, designated student or other resource person, followed by discussion;

- *Debate* - the pros and cons of a particular issue are argued by a team, and then the audience allocates their support for one opinion;
- *Student reports* - individuals post a report on which others then comment;
- *Discussion of set reading* - threaded discussions with comments on each required reading.

A range of studies has been conducted to test the effectiveness of these strategies on social aspects of the learning processes. Powers and Mitchell (1997) examined one graduate course which was offered entirely over the Internet. From the analysis of student performance and perceptions in this particular class, it is apparent that even as virtual classrooms become more prolific, the classroom community of learners can continue to flourish.

Another study investigated uses and gratifications of the web and it's social impact (Kaye 1998). The study identified six web use motivations - entertainment, social interaction, passing of time, escape, information, and web site preference.

A study of the attitudes of 29 graduate students enrolled in their first web-based class indicated a high degree of student-to-teacher and student-to-student interaction as well as accompanying high levels of satisfaction (Westbrook 1999).

One study examined a graduate distance course for elementary-school teachers using a "class web page" to increase teacher-student and peer interaction. Periodic student surveys during the course showed the dramatic impact of the class page on student interaction behaviours, technology use and literacy, and attitudes toward technology (Thompson et al 1998).

Our study addressed a different aspect of social interchange. There was no course designed on the net. Instead the study investigated the effects that a lecturer's personal web site can have on the relationships and online social community that can emerge from and follows a conventional academic course.

More and more instructors are building personal web sites and using them as teaching tools and as vehicles to achieving a variety of teaching purposes. Our study investigated the social role such a personal web site can exercise in enhancing the social links among the instructor's students following the usual social interchange which takes place in the classroom on an academic course.

To the best of our knowledge no other research study has been conducted into the influence an instructor's personal web site has had on the students' social interactions outside the walls of the classroom.

2. THE BAR-ILAN STUDY

2.1 Participants

The participants were 63 students takin one of the author's graduate courses at the University. During the acad mic school year, the students were given assignments via the instructor' site (http://www.passig.com) using a variety of Peer-To-Peer technologie . At the end of the academic year, the students were asked to answer a qu stionnaire in order to examine the influence of the web site on their extra cu icular social interactions.

2.2 P2P Technologies

We used different technologies to facilit e a variety of interchange and collaboration among the students. The three ain tools we used extensively were BrowseUp, eQuill and Groove.

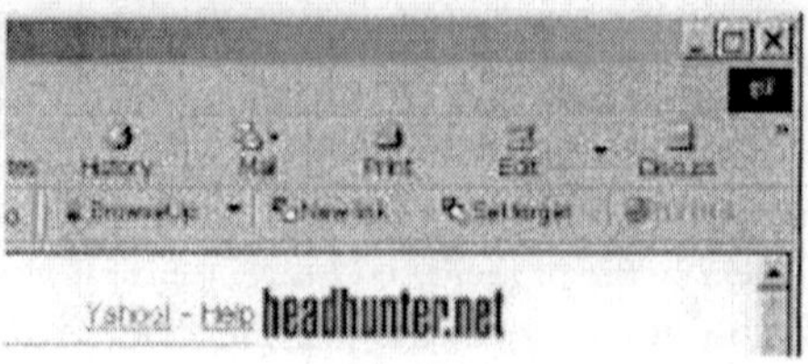

Figure 1. BrowseUp toolbar

BrowseUp (www.browseup.com) has at its core technology that enables users to create virtual links from existing online documents (Figure 1). Although these virtual links are inline links, they really reside on a link server (and not within the page), and the original document is not changed. These virtual links can point from both pictures and text within the page to associatively connected information like other URLs, the users local files (which are automatically published), and to new pages written on the fly by the user. Other users can view the virtual links while they access the URL from where the links point.

Figure 2. eQuill toolbar

eQuill (http://www.equill.com) was a nice tool (Figure 2). It streamlines and automates the manual process of designing, reviewing, approving, and maintaining Web sites. eQuill Markup Client allows you to mark up, highlight, edit, and post notes to Web pages or documents directly through your browser.

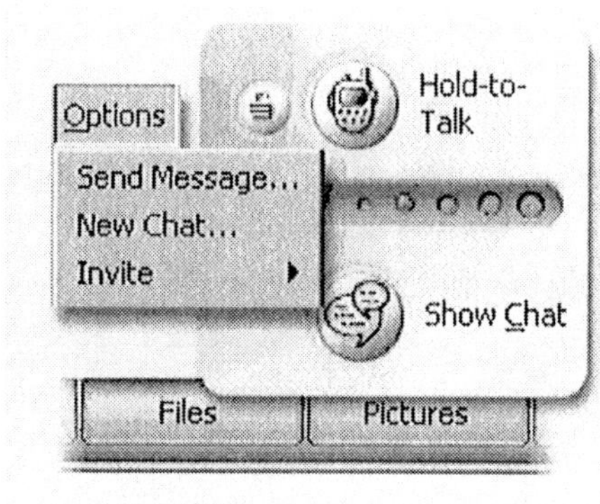

Figure 3. Some features of Groove

Groove was a collaboration tool we found to be very effective. It enables a small group of students to talk, work, play, and share things with team-mates inside shared spaces over the web (Figure 3). Groove lets you create interactive shared spaces as required, inviting the members you want and adding tools that match the task at hand. Groove complements existing productivity resources - including Microsoft Office and e-mail - with tools for communicating, sharing information and working together. You can see an example of a shared space at:

http://www.groove.net/products/individual/sharedspace.html.

Groove supports both online and offline use, letting the group work with each other in real-time or disconnected whenever they like, providing ongoing meeting and project management. The software works across network firewalls and other technical boundaries providing dispersed and mobile teams with on-demand, private networks.

Groove sits securely on the users PC, letting them control how, when and with whom they interact. No set-up configuration or server administration is required. After installing Groove, team-mates can hold meetings, edit presentations and write new ideas.

2.3 Forums and Students' list

We have provided a variety of topic-based forums to discuss issues raised during the class sessions. We have also posted a list of students with contact information. Only one student didn't want her name and contact information to be posted to the public through the site. The idea behind the students' list and the forums was to provide another means whereby social interaction can be enhanced.

3. EVALUATION

We couldn't find an appropriate questionnaire to evaluate the social interaction among students while using a personal web site. At the end of the academic year we asked the students to suggest a list of questions that can be circulated among the participants to evaluate the social impact of the lecturer's site on their social interchange outside the walls of the classroom. While this was not a viable validation technique, we found it suitable as an initial attempt to develop such a questionnaire. Through a collaborative effort we came up with a list of 14 statements. Participants were asked to evaluate the statements using a five-point scale (Table 1).

Table 1. List of statements

Statement	Strongly disagree (1, 2)	Undecided	Strongly agree (4, 5)
Site is a good vehicle to better socially connect the lecturer with the student.	-	-	78
Site a good vehicle to better socially connect the student with the lecturer.	-	-	76
The Forum is a good vehicle to get updated on the learning assignments.	52	12	36
I am an active participant in the Forum.	64	10	26
The Forum helped me communicate with other students.	64	10	26
I received suitable answers to my questions through the Forum.	67	13	20
The Forum can provide a suitable platform for open social interchange.	29	10	61
I use the students' list to locate contact information and e-mail addresses.	47	11	42
The P2P tools were used effectively through the lecturer's site.	35	35	30
The site is a good environment to serve the P2P tools and accommodate social interchange between students & lecturer.	30	16	54
The site is a good environment to serve the P2P tools and accommodate social interchange between the students.	20	25	55
The site with P2P tools adds additional social aspects to the learning process.	23	14	63
It's important to add P2P dimensions to academic courses.	17	9	74
The P2P experience was enjoyable.	12	16	72

4. DISCUSSION

In recent years there has been a growing pressure on academic and non-academic lecturers to develop and post personal web sites. The motive behind this pressure appears to be the promotion of the lecturer's professional image and the academic profile of their institutions. Most of the sites being developed aim to deliver personal information as well as a variety of teaching materials, assignments and complementary contents relevant to the interest and research scope of the lecturers. This trend is taking off among the vast majority of lecturers - worldwide and across disciplines. In just a few years it is reasonable to speculate that most academic lecturers will have a personal visibility on the web either through private servers or their institution's infrastructures. Personal web sites will probably take on the role of teaching tools for a variety of educational purposes. At this early stage it is important to develop new teaching strategies to better exploit the potential of personal web sites as teaching tools.

The underlying assumption of this study was that personal web sites could provide more than is normally thought possible today. Our assumption was that this educational tool is underestimated but it offers the opportunity to achieve educational aims that cannot be achieved in the conventional 'bricks and mortar' teaching environment. Indeed, the aim of this study was to investigate the role of a personal web site in providing new opportunities to further extend social interchange either around issues being discussed in classroom settings or around personal interchange.

The results clearly indicate that a personal web site with a variety of appropriate P2P technologies can further enhance the social extra-curricular interchange among students and lecturer as well as among students themselves.

One striking result of this study indicates that the students didn't see the Forums to be suitable for their social interchange. Most of the studies (Burstall 2000; Corbett and Eikum 2000) that looked at the role of Forums in the learning process did find some degree of learning interchange was achieved through Forums. However, it seems it did not provide social interchange, at least as it has been demonstrated in this study. One reason may be that of the nature of the Forum as a public venue, which was not the preferred option for conducting social and more personal and intimate interchange. Another reason may be that of the nature of the technology when set side by side with much more powerful P2P technologies such as Groove and BrowseUp.

Tomorrow's "brick and click" approach could provide new opportunities and challenges to teaching and learning. It would represent a convergence of old and new teaching techniques. The Internet is taking the basic bricks and mortar learning process - a physical location - and is moving it to the click of the Internet. This Internet presence has to be linked with the front end as well as the back end of Pedagogy. This study indicates that we need to reform our expectations of the new click and brick tools. On the one hand we need to lower our expectations from tools such as Forums and on the other hand raise our expectations in regard to P2P tools in the socialising process which accompanies learning.

REFERENCES

Burstall, J. (2000) Learning Communities for Social Change in Forums on the Web. *Australian Journal of Adult Learning*, Volume 40 (2) pp:33-52.

Corbett, D. R. and Eikum, D. (2000) *Social Impact of Distance Learning in Higher Education on Health, Physical Education, and Recreation Students.* Jones Institute for Educational Excellence, the Teachers College, Emporia State University, Emporia, KS 66801-5087. ED449160

Kaye, B. K. (1998) Uses and gratifications of the World Wide Web: From couch potato to Web potato. *New Jersey Journal of Communication.* 6 (1) pp. 21-40.

Kibby, M. (1999) *Teaching and learning on-line.* [http://www.newcastle.edu.au/department/so/interact.htm]

Moore, M. G. (1989) Editorial: three types of interaction. *The American Journal of Distance Education.* Volume 3 (2) Editorial.

Powers, S. M. and Mitchell, J. (1997) Student perceptions and performance in a virtual classroom environment. *Proceedings - Annual Meeting of the American Educational Research Association*, Chicago, Illinois.

Thompson, J.C., Malm, L.D., Malone, B.G., Nay, F.W., Oliver, B.E. and Saunders, N.G. (1998) Enhancing Classroom Interaction in Distance Education Utilizing the World Wide Web. *Mid-Western Educational Researcher.* 11 (4) pp. 3-7.

Tribe, D. (1994) An overview from higher education. In *Using Group Based Learning in Higher Education London*, L. Thorley and R. Gregory (eds.), Kogan. pp. 25-30.

Westbrook, T. S. (1999) Changes in student attitudes toward graduate instruction via Web-based delivery. *Journal of Continuing Higher Education.* Volume 47(2) pp. 32-38.

Potential problems of computer-mediated school education

Glenn Russell
Faculty of Education, Peninsula Campus, Monash University, PO Box 527, Frankston 3199, Victoria, Australia.

glenn.russell@education.monash.edu.au

Key words: Education, Virtual classroom, Values, Social Issues, e-learning

Abstract: The growth of on-line capabilities in school education has reflected the overall growth in society. However, there are both advantages and disadvantages. In the twenty-first century, the availability of on-line education increases the possibility that virtual experience will be substituted for reality. There are also concerns that there will be a blurring of appearance and reality, and that cultural imperialism will continue to spread through use of the World Wide Web. These factors are considered in terms of the need to establish future guidelines to reduce the adverse impact of the Web on school education.

1. INTRODUCTION

It has become a truism to state that there has been an exponential growth of on-line capabilities in recent years. Indeed, the growth of the Internet in commerce has been paralleled by rapid uptake by individuals, in their homes and elsewhere. This process has fostered the use of the Internet in school education and there has been a rush to wire classrooms, libraries and laboratories. It is likely that an increasing number of schools in developed countries will routinely use on-line components in their curriculum. Observers have noted that this adoption is sometimes associated with a shift from print-based materials and didactic teaching methods to a largely screen-based environment where the teacher acts more as a facilitator. However, such observations do not always take into account additional ways in which the Internet, and, in particular, the World Wide Web, affect students.

In this paper, I argue that several additional concerns can be identified and that guidelines may be required in the future if negative consequences are to be avoided. The problems I identify include the substitution of virtual experience for reality, the blurring of virtual and real, and the spread of cultural imperialism through use of the Web. These factors also operate in non-school contexts, and it is likely that, collectively, they are implicated in the ways that people understand the world around them.

2. IMPLICATIONS OF ON-LINE CAPABILITIES IN SCHOOL EDUCATION

2.1 The substitution of virtual experience for reality

Among the range of computer-mediated applications available for teachers to use with students are simulations. Merrill referred to these as being "a representation or model of a real or imagined specific object, system or phenomenon." Simulations can be valuable, particularly when the original experience is impossible to recreate for students for example a moon landing, an emergency in a nuclear power plant, or an historical event. However, as Turkle (1997) argues, there may be a problem if the simulation is inaccurate, as the students may never become aware of their lack of understanding. In a real-life situation, it may be necessary to discern a complex web of human behaviour and cultural viewpoints, and the programmer may have been unable to accurately translate such factors into the software. If software is deficient, takes a skilled teacher to recognise this and still teach effectively. An additional problem is that some teachers may use simulations for activities that previously did not require the use of a computer at all. For example, many senior high school students would have had the opportunity of being involved in the dissection of a frog or a rat in their science laboratory. It is possible to use a web-based simulation instead. However, it has been suggested that, although the use of virtual frogs would save frog's lives and be less messy, no two frogs are exactly alike (Armstrong and Casement 1998). The use of a simulation, in such an example, could lead to an imperfect understanding of frogs and biology.

The possibility that computer-mediated experiences can be limited when compared to real life can be overlooked. Multimedia generally incorporates images, text and sounds. As an experience it can be considered less rich when compared with a real-life context which also uses touch, taste and smell. Davy (1985) maintains the use of computer-based activities in place of traditional activities such as playground games can lead to "experiential impoverishment". This line of argument is not new. It has been argued that television replaces direct experience of the world with mediated experience and that the result was a lack of understanding of the real world (Mander 1978). In this respect, computer-mediated experience in the classroom exacerbates an existing situation whereby recognisable imperfections in a technology are accepted because the beneficial outcomes outweigh the disadvantages. However, such an attitude exposes people to unexpected outcomes and risks (Nissenbaum 1997).

Where the virtual experience substitutes for what teachers have often done in their classrooms, without clear evidence of it being educationally superior, there is room for suspicion of the motives of those involved. There may be a shortage of trained teachers or inadequate funding for laboratories and equipment. Underlying these factors are more serious questions about the commitment to teacher training and the funding of educational systems. Ironically, the decision to invest money in computers may have been at the expense of classroom supplies or resources. It is possible to imagine a hypothetical example of a school that decided to use virtual rats because of a shortfall in the science budget. If less money had been spent on computers and real rats had been available, the students might have developed better biological knowledge, and the cost may have been no greater.

When experiential education is subsumed by its computer-mediated counterpart, there is also the problem that students' abilities to form their own conclusions may be diminished - a possibility illustrated by a controversy at the National Museum of Australia (Shanahan 2001). The installation of high-tech exhibits in the indigenous gallery drew criticism because it was seen as politically correct. A more conventional exhibit might have displayed artifacts and allowed the viewers to form their own interpretations - the high-tech exhibit was seen as issuing fashionable interpretations. When exhibits do the interpretative work that individuals normally do there is an increased possibility of errors and false conclusions.

For schools, the implication of this example is that the importance of the teacher's role is increased. When students interact with a powerful medium which has the potential to inculcate the values of those who commissioned or programmed it, there is a need for teacher discussion so that students will be able to make informed decisions about the material. While this is a role which teachers have always undertaken, organisational decisions about teaching methods may run counter to it. In some cases, students may work in a library or resource centre without active teacher involvement; in others, they may work apart from the class using multimedia for enrichment or remediation. Teachers may also encourage homework that involves computer-mediated interaction and parents may not have the opportunity or skills to assist their children with the problems inherent in some computer-based materials.

2.2 The blurring of appearance and reality

Slouka (1995) maintains that we are surrounded by simulacra, in areas such as television and advertisements. Virtualisation in schools is likely to exacerbate an existing trend in society whereby students' learning about the world does not always come from a direct and real-life interaction with it. In Australia, a recent television advertisement for a CD-ROM featured a girl who claimed that she had lived among native peoples, had seen wild animals and had explored strange places. Her experiences had actually been based on her use of the CD-ROM. While adults would be aware that this was a marketing strategy, it would be easy to assume that the differences between virtual and real were unimportant. Young children might be able to understand that the virtual experience and reality were different, but they might also not fully appreciate the extent of that difference.

2.3 Computer-mediated experience and cultural imperialism

Cultural imperialism, in a technological context, refers to the potential for a technology to erode existing values and replace them with the values of a more dominant group or individual. When students use computers at home or at school the values of the software are transmitted to them. A student in Europe can access a clip-art gallery and find an over-representation of Americans in it, or a student in Japan can access the Web and find that most of the sites are in English. Some researchers consider the mind of the programmer is reflected in the form that the subject matter takes (Bowers 1988). In essence, the teacher surrenders some authority to the unseen programmer. While it can be argued that the teacher has always had to cope with bias in classroom materials, including books, the problem is more acute with computer-mediated materials.

There are three reasons why this may be so. The sheer volume of materials available through computers exceeds the ability of teachers to evaluate it. There is an increasing tendency for students to be asked to use computer-mediated materials at home, or in contexts where there is minimal opportunity for teachers to consider its implications. Finally, computer technology amplifies the power of individuals, through corporations, to influence others. Collectively, these factors can contribute to a classroom situation where teachers abrogate authority to an unseen programmer or the values of a corporation. The values of those involved in the programming and distribution of online materials challenge those of the classroom.

An emerging variation of this trend is advertising on web sites. In the pre-digital era, a student involved in a project on animals might simply have used a book to find the information. In contrast, a student in a classroom with web access might encounter a number of advertisements, the underlying values of which might run counter to those of teachers or parents. By concentrating too much on the enabling technology, and too little on the human factors involved in the educational process, computer-mediated learning using the web can result in a skewing of the values and understandings necessary for a developed society. There is, however, a contingent responsibility associated with the use of the web, illustrated by a reference to the introduction of motor cars. In the early twentieth century, these were also at a stage when their effect on society was not yet apparent.

3. IMPLICATIONS OF AN HISTORICAL PARALLEL

In 1901 a leading article in the London Times reflected on the growing popularity of the motorcar:

"The motor-car, as every one assures us, in phrase as new-fangled as itself, has come to stay. We cannot doubt it...We shall always have it with us ... we must all make the best of it. It is easy to draw an indictment against it...All the counts of such an indictment rest ultimately on the fact that the motor-car is new-fangled, unfamiliar and imperfect, like all human inventions at first. They have all been preferred from time immemorial against every new departure of human invention and progress...For good or evil, it will largely modify our social life."

The article went on to argue that drivers would need to exercise responsibility to avoid accidents and that authorities would have to consider registration and other regulations to avoid future problems. Similar comments could be made about the impact of the Internet on education. Although computers have had a place in school education for some years it is likely that the developments of the late twentieth century may yet prove to be only the first stage of a continuing involvement. In this context the growth of online educational applications may remind educators, students and parents that disadvantages as well as advantages can be expected

(Russell 2001). Consequently, it is appropriate to remind software developers, teachers and trainers that ethical choices must be made.

4. TOWARDS THE ETHICAL USE OF THE INTERNET IN SCHOOL EDUCATION

Discussions of ethical issues involved in the educational use of on-line computers often includes issues such as access to inappropriate sites, privacy and plagiarism. It can be argued that the issues discussed in this paper can be overlooked. Worldwide, parents, teachers and administrators use online computers on a daily basis. Sometimes they may be unaware there are choices to be made - awareness that virtual experience can be substituted for reality, understanding a blurring of appearance and reality is possible for learners, and an appreciation that predominant cultural mores on the Internet may conflict with those preferred in schools. Sometimes teachers who are confronted with these issues may elect not to use the Internet - at other times an awareness of the issues may help in the redesign of lessons that use on-line resources. To enable teachers to make these choices, teacher training and professional development will need revision. Educational authorities may need to consider guidelines to assist in this task.

REFERENCES

Armstrong, A. and Casement, C. (1998) *The Child and the Machine: Why Computers May Put our Children's Education at Risk.* Key Porter Books, Toronto, Ontario.

Bowers, C. A. (1988) *The Cultural Dimensions of Educational Computing: Understanding the non-neutrality of Technology. Teachers* College Press, New York.

Davy, J. (1985) Mindstorms in the Lamplight. In *The Computer in Education: A Critical Perspective*, D. Sloan, (ed.), Teachers College Press, New York, pp. 11-20.

Khalili, A. and Shashaani, L. (1994) The Effectiveness of Computer Applications: A Meta-Analysis. *Journal of Research on Computing in Education*, Vol. 27, No. 1, pp. 48-61.

Mander, J. (1978) *Four Arguments for the elimination of Television.* Morrow Quill, New York.

Merrill, P. F., Hammons, K., Vincent, B. R., Reynolds, P. L., Christensen, L. and Tolman, M. N (eds,). (1996) *Computer in Education.* Allyn and Bacon, Boston.

Nissenbaum, H. (1997) Accountability in a Computerised Society. In *Human Values and the Design of Computer Technology*, B. Friedman (ed), Cambridge University Press, Cambridge. pp. 41-64.

Russell, G. (2001) Computer Mediated School Education and the Web. *First Monday*, Vol. 6, No. 11 (November 2001).
[http://firstmonday.org/issues/issue6_11/russell/index.html]

Shanahan, A. (2001) An Interpretive dance across history. *The Australian*, Tuesday August 7, 2001, p. 15.

Slouka, M. (1995) *War of the Worlds: Cyberspace and the high-tech assault on reality.* Abacus, London.

The Times (London) (1901) Thursday September 5 1901, p. 7.

Turkle, S. (1997) Seeing through Computers: Education in a Culture of Simulation. *The American Prospect*, March-April, pp. 76-82.

Modern curriculum development for Informatics (Computing Science)

Tom J. van Weert and Fred Mulder
Expert Center for Educational Technology CETIS, Hogeschool van Utrecht, P.O. Box 85029, 3508 AA Utrecht, The Netherlands, t.vweert@cetis.hvu.nl

Executive Board, Open University of the Netherlands, P.O. Box 2960, 6401 DL Heerlen, The Netherlands, fred.mulder@ou.nl

Key words: Computer Science, Curriculum Policies, Higher Education, Responsibility, Society

Abstract: Modern curriculum development should fulfil specific requirements that reflect developments in society. A proposed set of requirements is compared against two recent curriculum development initiatives in the area of informatics (computing science). The two initiatives are the IEEE-CS/ACM Computing Curriculum 2001 (Joint Taskforce 2001) and the IFIP/UNESCO Informatics Curriculum Framework 2000 (ICF-2000) (Mulder and van Weert 2001). Detailed comparison showed that the principles used in these two initiatives cover the proposed requirements, although not one-to-one and with different emphasis. A difference in approach between the two curriculum initiatives, concerning the linking of societal needs and student competence development, is illustrated by explaining ICF-2000 Professional Categories and Graduate Profiles.

1. CURRICULUM DEVELOPMENT TRADITION

There is a long tradition in curriculum development in higher education. A common approach is to make a list of topics from the discipline involved, ordered by topic area and increasing complexity. Didactic guidelines are then added to specify how these topics should be taught to students. The curriculum effort is driven by the content of the discipline and many curriculum discussions focus on how important specific topics are to the curriculum. Clearly topic stakeholders play an important role in these discussions and may bias the resultant decisions.

Education exists to serve society: to supply society with capable graduates for its workforce and to assure common values. Mass education for industrial society did exactly that when it started, through its overt and hidden curriculum. However, mass education today has become a complex system with high inertia. It is more involved with its own, autonomous problems than with what is happening in society. Consequently many curriculum developers take more account of the needs of the educational system and its hierarchy than of the needs of society.

2. DEVELOPMENTS IN SOCIETY

We live in societies that are affected by the trend of globalisation, where English is becoming the “world Esperanto” and the job market increasingly global. Professionals may therefore be active in several different cultures and business settings.

Our societies exhibit a rapid rate of change and innovation is essential. Knowledge has become a common commodity; access to knowledge and reproduction of that knowledge is becoming easier and easier. On the other hand the development of knowledge has become a common activity of the new professional. "Time to market" is critical and therefore there is a need to develop new knowledge that is applied almost immediately in innovative situations. "Life Long Learning" is nothing special - it's now a characteristic of professional life.

Modern society experiences many problems that need a multi-disciplinary team approach and professionals work in such teams in which hierarchy is not important and respect is earned by achievement.

Clearly modern society is facing new types of problems. Their resolution requires a new type of professional who should be able to:

- Deal with these new types of problems in innovative ways.
- Create the new knowledge that must be applied to solve these problems.
- Work in multi-disciplinary teams and in an international context.

3. REQUIREMENTS FOR CURRICULUM DEVELOPMENT

The developments outlined in the preceding section have implications for higher education and curriculum development.

Modern society wants to know that what it gets is what it needs. Therefore a first, overall requirement for curriculum development is accountability. Curriculum developers have to answer the following questions:

- Why should the curriculum be developed?
- For whom is the curriculum primarily developed?
- What will the curriculum address, and why?
- How will the curriculum be realised in an effective way?

These key questions can be refined into requirements and are listed in Table 1.

Table 1. Requirements for modern curriculum development

Key Questions	Curriculum Requirements
Why should the curriculum be developed?	0 - Meet society's need to deal with new types of problems 1 - Meet society's need for professionals 2 - Meet need of individuals for personal development
For whom is the curriculum primarily developed?	3 - Students who develop competencies a. to build up a broad knowledge base b. to apply this base in a professional context c. to move along a path of personal growth
What should the curriculum address?	4 - Application of knowledge 5 - Core disciplinary competencies 6 - Professional competencies 7 - Personal growth achievements 8 - Interdisciplinary characteristics 9 - Globalisation effects
How should the curriculum be realised?	10 - Through a pedagogic and didactic approach 11 - In an effective and efficient process 12 - With implementation flexibility 13 - Combining sustainability and frequent updating

3.1 Informatics curriculum development in practice

At the 2001 International Federation for Information Processing (IFIP) World Conference on Computers in Education (WCCE2001), in Copenhagen, two curriculum documents were discussed.

- **Computing Curricula 2001 (CC2001)**, Steelman Draft (August 1, 2001), produced by the Joint Taskforce on Computing Curricula of the IEEE Computer Society (IEEE-CS) and the Association for Computing Machinery (ACM).
- **Informatics Curriculum Framework 2000 (ICF-2000)**, produced by Technical Committee 3 of IFIP on behalf of UNESCO.

Both curriculum documents are based on principles that can be interpreted as requirements for curriculum development.

3.2 Principles used to develop CC2001 (IEEE-CS/ACM)

The CC2001 Taskforce articulated the following principles (Joint Taskforce 2001)

1. Computing is a broad field that extends well beyond the boundaries of computer science.

2. Computer science draws its foundations from a wide variety of disciplines.

3. The rapid evolution of computer science requires an ongoing review of the corresponding curriculum.

4. Development of a computer science curriculum must be sensitive to changes in technology, new developments in pedagogy, and the importance of lifelong learning.

5. CC2001 must go beyond knowledge units to offer significant guidance in individual course design.

6. CC2001 should seek to identify the fundamental skills and knowledge that all computing students must possess.

7. The required body of knowledge must be made as small as possible.

8. CC2001 must strive to be international in scope.

9. The development of CC2001 must be broadly based.

10. CC2001 must include professional practice as an integral component of the undergraduate curriculum.

11. CC2001 must include discussions of strategies and tactics for implementation along with high-level recommendations.

3.3 Principles used to develop ICF-2000 (IFIP/UNESCO)

These principles were derived from the introduction of ICF-2000 (Mulder and van Weert 2000; van Weert 2000; Mulder and van Weert 2001)

1.The curriculum framework has an international scope and is developed for flexible and efficient implementation in different countries and different cultural and economic environments.

2. ICF-2000 takes a generic and inclusive view on the field of informatics.

3. ICF-2000 offers a global specification of necessary professional categories and profiles in terms of student competencies.

4. Professional categories are to be identified for which the curriculum should provide education; these categories should take account of other disciplines.

5. Graduate profiles are to be identified to cater for the educational needs of the professional categories in an efficient way; these profiles should take account of other disciplines.

6. Graduate profiles should build one upon the other allowing for efficient and flexible implementation in different educational environments.

7. Implementation of the curriculum should be possible within several resource situations.

8. Core informatics themes are to be identified and addressed in curriculum units; these themes should include personal and inter-personal skills.

9. Levels of competence should be specified for graduate profiles and curriculum units.

10. Learning approaches are to be specified for curriculum units.

11. International curriculum sources should be used for building an actual curriculum assuring that the curriculum is "state of the art", easy to maintain and allows for use of existing learning materials in many languages.

3.4 Requirements analysis

Comparison of the list of requirements shown in Table 1 with the principles used in the construction of the curriculum documents produces the result shown in Table 2. The classification is not straightforward and is open to debate. Also the authors are biased through their involvement in the development of ICF-2000. However, Table 2 illustrates that the requirements identified in the beginning of this paper are generally satisfied by the practical curriculum development for informatics (computing science).

Table 2. Coverage of requirements in practice

Requirements	**CC2001 Principles**	**ICF-2000 Principles**
Overall accountability	Yes	Yes
Why should the curriculum be developed?		
Meet the need of society to deal with new types of problems	4 (re. new technology)	1, 3, 4
Meet the need of society for professionals	9	3, 4 5
Meet the need of individuals for personal development	4	3, 5, 8
For whom is the curriculum primarily developed?		
Students who develop competencies		3, 9
a. to build up a broad knowledge base	1, 2	2, 3
b. to apply this base in a professional context	10	4
c. to move along a path of personal growth		3, 9
What should the curriculum address?		
Application of knowledge	10	3
Core disciplinary competencies	6, 7	8, 9
Professional competencies	10	3, 4, 5
Personal growth achievements		8, 9
Inter-disciplinary characteristics	2	4, 5
Globalisation effects	8	1
How should the curriculum be realised?		
Through a pedagogic and didactic approach	4, 5	10
In an effective and efficient process	5, 7	1, 5, 6
With implementation flexibility	11	1, 6, 7
Combining sustainability and frequent updating	3, 4	11

When the two lists of curriculum principles are compared, it can be noted that the principles of CC2001 are of a more general nature than the principles of ICF-2000. In the last case reference is made in Principles 3, 4 and 5 to the curriculum elements 'professional categories' and 'graduate profiles'. In fact the principles of ICF-2000 in part already explain how the actual curriculum development will meet some of the principles.

The comparison also shows that in Principle 3 of ICF-2000 student competencies are introduced as an educational aim (to be realised in graduate profiles and curriculum units) while in CC2001 student competencies are only referred to in Principle 10. The combination of observations offers an explanation for the appearance of ICF-2000 Principles 3, 4 and 5 in several places in Table 2.

4. COMPARISON OF CC2001 AND ICF-2000

Comparison of the principles of the curriculum documents using the requirement framework in Table 2 shows that, in the case of ICF-2000, more attention appears to be given to the following requirements:

1. Meet the need of society for professionals
2. Meet the need of individuals for personal development
3. Students who develop competencies ... (c) to move along a path of personal growth
7. Personal growth achievements
8. Inter-disciplinarity characteristics

A more in-depth study of the documents shows that there indeed is a difference. ICF-2000 (Mulder and van Weert 2000 - Sections 4 and 5) tries to specifically link - *societal needs*, to *categories of professionals*, to *competence development of students* - before moving to a curriculum description (Mulder and van Weert 2000 - Sections 9 to 12).

Such a link is more diffuse in CC2001 (Joint Taskforce 2001) which centres on a topical curriculum description in Chapter 5 and gives a "variety of approaches to cover core units in the body of knowledge" in Chapter 7 (Introductory courses) and Chapter 8 (Intermediate courses). It only refers to Professional Practice and Characteristics of Computer Science graduates in Chapters 10 and 11. In view of this difference in approach the specific link between societal needs and competence development of students merits separate attention.

4.1 Linking society needs and student competence development

ICF-2000 deals with university and higher professional education. It gives a specification of categories of professionals in section 4. What informatics competencies do professionals, who have been educated in higher education and are starting their professional career, need? This knowledge should fit the role these professionals will fulfil in economic life, in an effective and efficient way.

4.2 Three main categories

Looking at the role professionals play in the economic process three categories of professionals can be identified, acting or interacting with Informatics (I) in a broad sense:

A I-users:

Non-I-professionals using ready made I-technology or I-applications in their work, for example non-informatics professionals using digital communication systems or text processors. Many professionals in highly developed economies are I-users.

B I-appliers:

Non-I-professionals applying I-knowledge and I-skills in areas different from informatics, for example non-informatics professionals in the economic field who use I-applications to model economic situations.

C I-workers:

I-professionals working in the field of informatics like the system engineer who develops I-applications or I-technology (note that the field of informatics is broad with diffuse boundaries with other disciplines).

It is important to note that I-appliers and I-workers generally will also be I-users, while I-workers may or may not be I-appliers.

4.3 Categories of professionals

Within the main categories eight subcategories are identified.

A1 Instrumental I-users

They use computer technology or software packages in their work. Examples can be found in word processing, using databases, making spreadsheets, preparing presentations, graphical drawing, communicating by e-mail, retrieving information through the Internet and videoconferencing.

B1 Conceptual I-appliers

Conceptual I-appliers apply specific knowledge from the informatics domain or typical informatics skills in their own area that is increasingly infused by informatics. Examples include teachers who design computer supported education, chemists doing molecular modelling, computer artists, the media specialists applying internet technology and economists introducing electronic commerce.

B2 Interfacing I-appliers

These professionals combine knowledge and skills from their own area or profession with informatics knowledge and skills, in an interfacing role linked to I-professionals. Examples are the business consultant advising on the development of computer based information systems, the mechanical engineer participating in industrial automation projects, the sociologist supporting the introduction of large scale work computerisation, the ethicist advising on privacy matters and the lawyer collaborating on the formulation of software contracts.

B3 Researching I-appliers

Researching I-appliers combine knowledge and skills from their own research areas with informatics knowledge and skills in research efforts that connect and integrate informatics with other disciplines. In general interdisciplinary research projects show fruitful collaborations with informatics researchers (C3) and are designed to open up new horizons. Examples can be found in research on distance learning, human-computer interaction, cognition, computational science, telematics & multimedia, linguistics, information science, knowledge technology and logistics.

B4 Directing I-appliers

Directing I-appliers are well equipped and skilled through their general understanding and broad overview of informatics, to play a policymaking, supervisory or managerial role in the areas of I-technology and I-applications. The project manager running a project in which new I-technologies are introduced into an organisation or in which a tailored I-application is being developed is one example. Others include the information (policy) manager supervising an organisation's information processing and the communication (policy) manager responsible for creating, updating and exploiting an Internet site of substantial volume.

C1 Operational I-workers

They have a thorough understanding of and well-developed skills in informatics as a broad discipline, more specifically in the area of exploitation, control and maintenance of available I-technology and I-applications. Clearly this category contains a large portion of lower level I-professionals, for example computer operators, network operators, application administrators, database administrators and helpdesk employees. University-level I-professionals will be required in directing, supervising and managing roles in this area (note a possible overlap with B4).

C2 Engineering I-workers

Engineering I-workers have a thorough understanding of and well-developed skills in informatics as a broad discipline, more specifically in the area of analysis, design and implementation of I-systems. Examples include the information systems analyst, the software engineer, the knowledge engineer, the scientific programmer, the database developer and the integrated circuit designer.

C3 Researching I-workers

Researching I-workers have a thorough understanding of and well-developed skills in informatics as a broad discipline, more specifically in research. They are supposed to further develop the I-discipline and its concepts, both on its own and in relation with other disciplines, in the latter case collaborating with researchers from category B3.

5. GRADUATE PROFILES

To allow effective education of all eight professional categories ICF-2000 identifies the following four graduate profiles (Mulder and van Weert 2000 - Section 5) that build upon each other.

	Professional ICF-2000 Category	Orientation
1. **BIP**	A1 [Section 9]	Awareness
Basic Instrumental I-Profile		
2. **BCP**	B1 [Section 10]	Application
Basic Conceptual I-Profile		
3. **MIP**	B2, B3, B4 [Section 11]	Design
MInor I-Profile		Modelling
4. **MAP**	C1, C2, C3 [Section 12]	Conceptualisation
MAjor I-Profile		Abstraction

5.1 Profile fingerprint and orientation

The professional categories grouped in a graduate profile have characteristics that reflect differences between the graduate profiles. These differences are related to:

- The coverage of informatics core curriculum themes which 'fingerprints' the graduate profile for a specific category of professionals.
- The orientation in terms of goals and competencies to be reached.

Core curriculum themes

Twelve I-curriculum themes are identified that globally offer a complete coverage of the informatics field. The themes were developed at an IFIP Working Conference (Mulder and van Weert 1998) and essentially set the curriculum framework which defines the core of informatics education. These also allow the relationship of informatics with other disciplines to be clarified.

1. Representation of information;
2. Formalism in information processing;
3. Information modelling;
4. Algorithmics;
5. System design;

6.Software development;
7.Potentials/limitations of computing and related technologies;
8.Computer systems and architectures;
9.Computer-based communication;
10. Social and ethical implications;
11. Personal and interpersonal skills;
12. Broader perspectives/context (links to other disciplines).

The coverage of the themes in a graduate profile constitutes a 'fingerprint' fitting specific categories of professionals.

Orientations

Each graduate profile has a dominant orientation in term of goals with competencies fitting specific categories of professionals. There are four orientations:

AW AWareness (know or use)

Aimed at developing basic knowledge and skills that allow students to acquire basic literacy with respect to informatics in general and to perform standard operations using computer technology or software packages;

AP APplication

Aimed at developing basic conceptual understanding of informatics and of some more advanced informatics skills which allow students to apply basic informatics to other disciplines or areas;

DM Design and Modelling

Aimed at developing a general understanding and broad overview of informatics, especially with respect to the modelling and design of informatics applications;

CA Conceptualisation and Abstraction

Aimed at developing a thorough understanding of, and well-developed skills in, informatics as a broad discipline, the essence being to further develop the capability of students to abstract and to conceptualise.

REFERENCES

The Joint Taskforce on Computing Curricula (2001) *Computing Curricula 2001*. Steelman Draft (August 1, 2001. IEEE Computer Society & Association for Computing Machinery, IEEE CS/ACM.

[http://www.acm.org]

Mulder, F. and van Weert, T. J. (2000) *Informatics Curriculum Framework 2000 for higher education [ICF-2000]*, UNESCO, Paris.

[http://www.ifip.or.at/pdf/ICF2001.pdf]

van Weert, T. J. (2000) Guidelines for Authors, *Informatics Curriculum Framework 2000 for higher education [ICF-2000]*, UNESCO, Paris.

Mulder, F. and van Weert, T. J. (2001) *IFIP/UNESCO's Informatics Curriculum Framework 2000 for Higher Education*, SIGCSE Bulletin, Association for Computing Machinery (ACM), Vol 33, No4, December. pp. 75-83.

Mulder, F. and van Weert, T.J. (eds) (1998) *Informatics in higher education: Views on informatics and noninformatics curricula*, Proceedings of the IFIP/WG3.2 Working Conference on Informatics (computer science) as a discipline and in other disciplines: what is in common? Chapman & Hall, London.

Innovative pedagogical practices using ICT - results of the German SITES-M2

Rebekka Dalmer, Thomas Petzel and Renate Schulz-Zander
Institute for School Development Research, University of Dortmund, Vogelpothsweg 78, D-44221 Dortmund, Germany

schuza@ifs.uni-dortmund.de

Key words: Classroom Teaching/Practice, Innovation, Research, Students, Teacher

Abstract: The international SITES-Module 2 (Second Information Technology in Education Study) of the International Association for the Evaluation of Educational Achievement (IEA) is a qualitative study of innovative pedagogical practices that use Technology in schools in 28 countries. The investigation was carried out in 2000/2001. Twelve case studies have been conducted in primary, lower and secondary schools in Germany. Results of the German SITES-M2 concerning the outcome of teacher and student practices are presented and discussed in the paper.

1. INTRODUCTION

The international study Second Information Technology in Education Study (SITES), of the International Association for the Evaluation of Educational Achievement (IEA), covers a research programme comprising three modules (Pelgrum and Anderson 1999). SITES-Module 2 is a qualitative study that aims at identifying, analysing and documenting innovative forms of teaching and learning worldwide that include the use of ICT and where that use is outstanding. Case-studies on innovative pedagogical practices with the use of ICT were carried out in schools in 28 countries in 2000/2001. The results supply decision makers and teachers with exemplary teaching practice illustrating the use of ICT and with findings on the factors which are necessary for the successful implementation of ICT and its sustainability. They contribute to the development of educational theory and will assist the implementation of SITES-M3. This paper presents selected results from the German SITES-M2.

2. RESULTS OF THE GERMAN SITES-M2

For the evaluation of the comprehensive results, the international analysis scheme was used. The system of categories stipulated by this scheme was applied for each individual case. The analysis scheme covers different levels of analysis. The core field of the case reports, and therefore also of the cross-case analysis, relates to the so-called micro-level: the specific activities of the teachers and students during the implementation stage of the innovation. First of all the essential characteristics of the activities of the people involved are described. Then the effect of the innovation on both the students and the teachers is examined to identify how those involved profit from the innovation. Problems and obstacles that may occur during the implementation phase of the innovation, and estimates concerning the sustainability of the innovation, are given. For example we wish to identify the conditions which have to be fulfilled to integrate innovative uses of ICT into teaching which are sustainable.

2.1 Teacher Practices and Outcomes

Questions such as which specific activities do teachers carry out during the innovation and what affect does the innovation have on those involved are dealt with in this category. 'Teacher practice' covers both the preparation of the teacher for the innovative practice and the actual activities carried out during the realization of the lesson. During the period of innovation, the teacher takes up an advisory, guiding or instructing role and also provides the structure for the activities of the students. The interviewees involved in all the case studies either mention these two characteristics of teaching activity or the characteristics could be observed during the visits to see the lessons (Table 1).

"The idea is that we organise, i.e. provide cause for learning and organise its framework conditions. The impetus is then given and they work until they are through and we proceed to our next duty of evaluating the whole work. In-between, we are helpers when consulted." (Teacher DE003)

Table 1. Teacher Activities

Activity	Total Case (12)	Cases - Primary (4)	Cases - Lower Secondary (4)	Cases – Upper Secondary (4)
Lecture	11	3	4	4
Advise or guide students	12	4	4	4
Create structure for student activities	12	4	4	4
Design and prepare instructional material	9	4	2	3
Monitor and/or assess student performance	9	3	2	4
Collaborate with students	6	3	2	1
Collaborate with colleagues	7	2	2	3
Collaborate with external actors	5	1	1	3

In 11 of 12 cases the teachers adopted, at some time, the classic role of the teacher, imparting knowledge in the traditional ex-cathedra manner. Ex-cathedra teaching was considered necessary, especially in the introductory phases of an innovative lesson to offer a starting point for the teachers' work with the students. In the course of the lesson there are always teacher-centred phases in which the results of the students' work are being collected and new tasks are set for the students. The following quotation illustrates the two poles of teacher activity - ex-cathedra teaching on the one hand and the adoption of a more restraining, background role that gives enough freedom for independent work to the students on the other:

"Frontal teaching is inevitable, it must be introduced. Else, I will try to make the children work independently and as much as possible always with assistance. In every teaching form, there are always children who feel at home with the method and others who have difficulties." (Technical Coordinator DE009)

If you distinguish the different activities of the teachers by grades, a more differentiated picture may be discerned (Table 1). In most cases teaching activity includes the 'design and preparation of instructional material'. Many of the teachers who were interviewed pointed out that there is no instructional material available for their innovative work with ICT that they can fall back on. In the majority of cases the teachers said that the checking or rating of student performance was a further substantial activity. Changes in cooperation of teachers with students were explicitly mentioned in six cases and seem to be a characteristic of innovation, particularly in elementary schools (in 3 cases). With regard to the case studies in upper secondary, cooperation with actors outside of the classroom seems to be an essential characteristic of the changes that have taken place.

To sum up it may be said that in all cases, the teachers adopt an advisory and guiding role in their teaching and that they are responsible for the structuring of student activities. In the observed innovative pedagogical practices, phases of teacher-centred practice take turns with phases that promote (and require) greater student independence. In the latter phase teachers withdraw themselves more and the students control their own learning and work-processes.

Table 2. Impacts of innovation on teachers

Impacts	Number of Cases	Cases - Primary	Cases - Lower Secondary	Cases – Upper Secondary
New pedagogical skills	11	3	4	4
ICT skills	7	2	3	2
Collaborative skills	4	1	1	2
Positive attitudes towards the teaching profession	7	2	3	2
Negative outcomes	5	2	1	2

The impact of the innovation on the teachers can be divided into different fields (Table 2). In connection with the realized innovations, most of those interviewed reported that they acquired new skills, pedagogical as well as technical, and experienced a change of attitude towards the events in teaching. On the basis of the given data it is not possible, though, to comment on the causality of this process - are the acquisition of new skills or attitudes an impact of the innovations or a prerequisite for their realization? Equally it was not possible to conclude that it was a reciprocal process of mutual influence between concrete practice and new skills.

In almost all cases it was reported that the innovation fostered the acquisition of new pedagogical skills. The teachers were able to teach in a project-oriented and open way.

"I am relatively new to the school and did not lay such emphasis on project-orientated lessons. The amount of lessons that now take on this form has risen due to the use of computers in the classroom. I have developed a new style of teaching. I now have the possibility to go about it in a different way. You have to go through this inner process and think in a different way." (Teacher DE006)

The teachers have learned to handle their new role as an 'advisor' and 'tutor'.

"[One has to] adjust himself to the new role of the teacher. The [traditional role of teachers] is a different one. It is such with a tendency of constantly standing behind the student to see what he is doing. And when one is then sitting there like I was and no one wants anything from you, it is somehow awkward. One feels useless. One simply has to stand the situation." (Teacher DE015)

A new teaching situation develops from this. Consequently teachers have had to adapt to changed requirements of their role. They must be able to handle new situations in their classrooms.

"That is a very intimate pedagogical situation, where you do not always need the teacher as an authority. I have learned very much. .... Today I am much more moderate in my references. In the past it was: "You cannot do it this way! This is totally wrong!" Today I would say: "Just think about it again - do you think that it is right this way?" Today I am also much more flexible and can think in terms of alternatives." (Teacher DE015)

In only one case the realized innovation was not accompanied by the acquisition of new pedagogical skills. This can be traced back to a ceiling-effect: the teachers in this school had been working with open or project-oriented forms of teaching for quite some time. Consequently the innovative work did not lead to the acquisition of new pedagogical skills.

Similarly, the acquisition of ICT skills as a concomitant of the innovation was mentioned in seven cases. In the remaining five cases knowledge of the use of ICT existed already so that the teaching activities could not increase it.

In five cases the teachers reported negative outcomes from innovation. In the first place, greater time and work expenditure was mentioned in the context of innovative work. Apart from creating new instructional materials it was also found extremely time-consuming to develop the use of ICT to integrate it innovatively into teaching. Some interviewed teachers identified a further negative outcome in the fact that due to the trial status that they are in as they are trying something new, they feel that it is more than they can handle. In addition, the realization of the innovative pedagogical approach requires more concentration from the teacher. The empirical findings within the scope of the evaluation of the German initiative "Schulen ans Netz" (Schools Online), carried out by the Institute of School Development Research at the University of Dortmund in 1999, show teachers noted that teaching itself is felt to be more demanding although it is more fun

(Huneshagen et al 2000; Schulz-Zander 2001).

2.2 Student Practices and Outcomes

In this section student activities during the period of innovation, as well as the impacts of the innovation on the students, are characterized (Table 3).

Table 3. Student activities

Activity	Total Cases	Cases - Primary	Cases - Lower Secondary	Cases - Upper Secondary
Drill and practice tasks	1	1	0	0
Research projects	3	0	2	1
Search for information	9	2	4	3
Solve problems	1	0	0	1
Manipulate/interpret data	2	0	0	2
Publish/present results	10	3	4	3
Design and create products	11	4	4	3
Collaborate	12	4	4	4
Collaborate with external actors	6	2	2	2
Assess own/peer performance	5	1	2	2
Select own tasks	9	3	4	2

The cooperation between the students was considered an important characteristic. The students mainly worked together in small groups or with a partner. The organisation of group work in the innovation class is described:

"We should undertake more with each other. In the past, we mostly, had individual work. Now, everything is done in groups: a reason for everyone to get along with each other. Table groups are always put together randomly. Lots are cast. Thereafter, one has to get along with them and do group work with them for three months. We do not do group work only with table groups, but also with them. The groups should be able to get along in this case." (Student DE003)

Much student activity consisted of designing and creating products as well as in publishing or presenting their results (PowerPoint presentation, posters, fairy tale book, homepage, newspaper).

In many cases, students searched for information - either in electronic media (Internet, CD-ROM) or print media (books, magazines). In most cases, students had the opportunity to at least partially have a say in the assignments they were set. The students were able to choose from a bigger collection of possible assignments.

With regard to the effects of the innovations on the students it was found that skills of cooperative working were promoted in almost all cases (Table 4):

Table 4. Impacts of innovation on students

Impacts	Total Cases	Cases Primary	Cases - Lower Secondary	Cases – Upper Secondary
New subject matter	7	2	2	3
ICT skills	12	4	4	4
Communication skills	7	2	3	2
Problem solving skills	1	0	0	1
Information handling skills	8	2	4	2
Team/collaborative skills	11	4	3	4
Metacognitive skills	8	3	3	2
Increased motivation	10	4	4	2
Outcomes for different groups	5	1	2	2
Negative outcomes	3	1	1	1

"Last year, the students learnt to work in groups. .... The students work together in a proper and astoundingly disciplined manner. Yesterday when I came too late, the students were seated and had started working. Under normal circumstances, no one does that in the eighth class. .... It raises the degree of self-dependability. .... They agree on appointments for meeting at home or simply begin to work." (Teacher DE003)

The independence of the students increased considerably due to the innovative work. They were increasingly able to control their learning process. In two thirds of the cases it was explicitly stated that the development of meta-cognitive skills or self-learning competences was promoted by the innovations.

It was also reported that in most cases the students worked and learned with increased motivation as a result of the innovation. This is shown by the following quotation. It also illustrates the way that the use of ICT affects lower-performing students (according to the opinion of some teachers):

"Especially children with extremely low performance took great profit out of working with the computer. Children who were almost functionally illiterate devotedly wrote with the computer. Motivation is extremely important here. They suddenly liked to do things they had actually hated before and they had better results, too. .... I think this is because they can always revise their work until it is good and rounded." (Technical Coordinator DE007)

In all the cases, the interviewees reported that the students improved their skills in the use of ICT. The following quotation not only comments on the acquisition of ICT skills but also on the development of information handling skills:

"Media competence is for me the active participation in this media society. Not only as consumer, also as producer. That's why I produce with the students whatever I can. In this way, they learn that is it possible to do a radio broadcast, or a video. They now look at these things in a different way. They suddenly see camera angles and say: "That is so cheap, that hasn't happened, this has been made without effort"." (Teacher DE013)

Among the positive impacts, a few negative outcomes were described. Fear was expressed that the innovative practice with ICT would reinforce differences in performance - both between students that have access to ICT at home and those without access at home on the one hand, and between students who were low or high performing even before the innovation began:

"For me there is a problem with a widening gap. You know, it's widening because if somebody has access to the net they do that really fast. Somebody who can't do it in the first place won't have the guts to sit down and try". (Teacher DE002)

"There is a difference between students who are good and those that are weak at learning. Here however, the fact already known by us about the open and closed forms of teaching is reproduced, namely that students who are apparently good at learning, motivated and highly cognitively qualified, are more at home with all forms of independent demands than those that are weak at learning and require more guidance. Since I celebrate the usage of computer as a means of opening up, it may be possible that they indeed, suffer a setback if one looses sight of this fact." (School Principal DE002)

Concern was also expressed about the "open-teaching" and how students experiencing this appeared to lack orientation.

3. PROBLEMS AND SOLUTIONS

The innovative pedagogical practices are often accompanied by problems. These problems, though, were, at least partly, resolved by the schools we investigated. In the first place these problems relate to the field of technical infrastructure. On one hand some schools find it difficult to finance the necessary hardware (4 cases) and this in turn is closely connected to the resulting low level of technical equipment (3 cases). Insufficient support by the supervisory school authority was criticised in many cases although due to the sponsorship raising activities of the principal it was possible to find mezzanine money. On the other hand technical support was considered a problem in four cases. There are few technicians who are responsible for the administrative matters associated with equipment in schools so this task falls to the teachers. Some schools have solved the problem by employing technicians.

A further set of problems affects all teachers. The burden of time to devote to innovative work was considered a problem in 4 cases. In 5 cases the teachers who were not involved in the innovation were critical at the beginning of the programme. They were envious of the equipment supplied to the innovation classes and annoyed that they had to teach extra lessons, for example when the innovation teachers were absent because of European school exchanges or attending coordination meetings. The teaching staff also expressed some scepticism of the innovation.

A further difficulty in three reported cases was that the timetables were too cramped. According to the interviewees, the 45-minute-rhythm of the timetable prevented the self-controlled work process of the students. On top of this, time restraints made multidisciplinary work more difficult. In one school this led to the establishment of a weekly project day where the usual division into several lessons was removed. This project day is particularly used for multidisciplinary work on projects. It is also used for more complex projects in individual subjects. Organisational problems were not always resolved but they did not prevent the sustainability of the innovative teaching practices.

4. CONCLUSION - SUSTAINABILITY OF THE INNOVATIONS

The data and findings that were gained within the scope of the case studies led to the conclusion that the sustainability of innovations mainly depends on the commitment of the involved teachers to continue the innovative activity. In one school it was considered important that there was a teacher who was most involved in the project and who acted as a driving force for the other teachers who were involved. According to two cases, the further training of the teaching staff in order to expand knowledge of innovative teaching techniques and of ICT is also important for this commitment.

Interviewed teachers also commented that it was important for the sustainability of the innovations that support for the innovation from the principal is guaranteed (3 cases), that innovation must be supported by all the teaching staff (3 cases) and that innovative work should be accepted by the parents (one case). The codification of the innovative teaching practices into in-school curricular stipulations, as it was expressed in two cases, have the most lasting effect on the innovative activities in schools and ensure further implementation of the innovations.

However, generalising statements can not be made on the basis of the available data. There is further need for research. The Survey SITES-Module 3 will build upon the results obtained from the first two SITES modules. The distinguishing components of SITES-Module 3 are a school/teacher survey and an optional student survey. Teachers will be asked about their attitudes, training and past experiences in respect of technology, and their classroom practices. The student survey will include indicators of knowledge, attitudes, exposure to different learning applications, and the use of ICT in the home (see http://www.sitesm2.org).

REFERENCES

Hunneshagen, H., Schulz-Zander, R. and Weinreich, F. (2000) Schulen am Netz: Veränderung von Lehr- und Lernprozessen durch den Einsatz neuer Medien. In *Jahrbuch der Schulentwicklung*. Band 11. Weinheim, pp. 155-180.

Pelgrum, W. J., Anderson, R. E. (eds.) *ICT and the emerging paradigm for life-long learning*. IEA, Amsterdam.

Schulz-Zander, R. (2001) Using New Media for Learning in Schools. In *Futures of Education. Essays from an Interdisciplinary Symposium*. Bern. pp. 83-101.

e-learning@alma-mater.de - Net-based distance education in the traditional university

Paul-Th. Kandzia
Institut für Informatik, Albert-Ludwigs-Universität Freiburg, Georges-Köhler-Allee, D-79110 Freiburg

kandzia@informatik.uni-freiburg.de

Key words: Distance Education, Tele-Teaching, Virtual University

Abstract: E-learning will not inevitably cause a revolution of the traditional university. On the contrary, it is not yet clear whether multimedia and network technology will find its place in the university to benefit or profit the students. We discuss various methods and techniques (including Learning Management Systems, synchronous tele-teaching and Web-based seminars) and propose best practices, which can succeed under the specific conditions of a traditional university.

1. INTRODUCTION

Not that long ago Encarnacao et al (1999) predicted the end of the on-campus university as we know it by the year 2005. They maintained that E-learning will become the common form of education, educational providers will operate world-wide and every university will have to fight hard for its position in the global market. Only a few élitist universities will maintain classrooms in which excellent teachers still meet their students face-to-face.

Whether this scenario is a promise or a threat has been discussed passionately since the paper was published. A lot of activity has been generated. Federal and state programs almost everywhere have fostered hundreds of ambitious projects in Germany and Switzerland under the heading of "Virtual Universities XY".

However, in the Europe of the year 2002 the traditional university is still very much alive, unwilling to give up its place to its virtual counterpart. There are strong reasons why this will not change in the next decade. First, distance education is not desired as the general model of studying, either by the universities, or by prospective students. The drop-off rate is significantly higher than in traditional studying, and many people still appreciate campus-life as an important phase of personal development. Secondly, even if universities would like to develop net-based distance programmes on a larger scale, they can't afford it. For the production of just two masters programmes in the *Virtual University of Applied Sciences* over 20M Euro were needed. The *Virtual Campus Switzerland* (a federal program, not an institution) spent more than 100,000 Euro in developing a course of 3 ECTS credits. These figures don't include the costs of actually running the courses.

Clearly, a virtual revolution of the alma mater is not in sight. Rather, the question is how universities can take advantage of "new media" technology. Indeed, we face the great danger that most of the results from good projects are not adopted into everyday activity and that the great political, scientific and financial effort may not give rise to any substantial new possibilities for students.

Through our involvement in several projects we have gained a lot of experience with different forms of web and multimedia based education (Kandzia and Ottmann 2001). In this paper their usefulness for a traditional university will be discussed. In particular, we stress the following criteria. First of all, an e-learning method or technique must produce added-value for teachers or students. In most cases this will be more time and space flexibility without a loss of learning success, but it may be a gain of pedagogical quality as well.

Further, if this scenario is to be used in the everyday-life of a traditional university, it has to fit smoothly into normal university work, tools which are easy-to-use have to be available, and it should not be too costly.

2. E-LEARNING OPTIONS

2.1 Learning Management Systems (LMS)

A Learning Management System (LMS) provides the basic functions needed for composition and delivery of any web-based educational offering administration of users and courses, composition and delivery of content, communication, assessment and grading. As a central service an LMS must be installed and run by a central department within the university. Leaving this task to single small projects is extremely inefficient especially if different projects run their courses on different platforms that are not interoperable (a situation found in many universities).

In most cases it is unreasonable for a university to develop its own LMS. There are many commercial solutions on the market, some of them open-source, and even when their drawbacks are taken into account, universities will find it difficult to provide a better alternative and support it over many years.

2.2 Synchronous Tele-teaching

In synchronous scenarios a live lecture or seminar is transmitted to one or more participants using video, audio, and a shared whiteboard. In a distributed seminar the teacher's work is much more involved than in one classroom alone: he/she has to motivate group work and participation to compensate for the loss of face-to-face communication, e.g. by common presentations of two students from different locations. Students also require special training to use tools like a shared whiteboard. However, those skills are very useful and appreciated by the students.

The main drawback of this scenario is that it only yields satisfactory results if the teachers of all connected sites work together very closely, since co-ordination of timetables, curricula, and credits is essential. Furthermore, special rooms with data projectors and advanced input devices like graphics tablets as well as high-speed connections are needed - we use MPEG2-Hardware and a virtual LAN over ATM. There should also be technical support. In home learning variants video quality is still poor. All in all, synchronous tele-teaching does not satisfy the goals of time and space flexibility. Although in certain cases with the right partners and thorough planning it may yield a convincing performance it cannot be generally recommended.

2.3 Web-based Seminars

In contrast to seminars based on a form of videoconferencing, web-based seminars mainly rely on textual communication - newsgroups, mail and chat. Material in normal web-formats is shared easily. This scenario is cheap and, from a technical viewpoint, is easy to use. It is fully flexible in terms of time and can be used from the student's home PC. The organisational effort required on the university side is quite limited. Web-based seminars are found in the philological, political and social sciences - all of which involve working with texts. However, it would appear teachers of other disciplines consider the pedagogical potential is too restricted. A very appealing didactical feature of web-based seminars is that professionals from outside the university can contact and work with students very easily.

2.4 Courseware Programming

With courseware (or WBT) programming we designate the exclusive authoring and design of web-based material. The end result may consist of simple HTML, but usually audio, images and animations are used. In high-end productions content exposition and the more interactive elements of self-tests, assessments or simulations are closely interleaved. Sometimes courses are adaptable to the learners' progress, for example, by making the material that is presented dependent on test results. Clearly the pedagogical potential is exciting and hence it is not surprising that the majority of "Virtual University" projects aim to produce courses of this type. However, the costs of courseware programming (in terms of time, manpower, and money) are extreme and are frequently underestimated, especially if the product goes beyond HTML and the didactic potential is really exploited (Rumble 2001). If a project is already over-stretched by producing static content, it will never be capable of giving appropriate student support or providing regular course updates. As a result, many ambitious university projects only produce a kind of digital textbook and never lead to courses students really profit from. The effort never achieves "pay back".

In special cases it may be reasonable and affordable to develop high-end WBT modules. Factors that justify such development include subject matter that is difficult to understand when taught in a traditional way, where the subject matter doesn't have to be revised frequently or where the WBT is used by large numbers of students. For example, when Java-Applets were used to visualise discrete cosine transformations the learning results were better than those from lectures. Problem-based-learning with a "virtual patient" in Medicine has also proved very successful.

In addition to the enormous time and money needed there is often a structural problem - courseware programming does not fit smoothly into the organisation of a (Humboldt) university. Universities mainly employ scientific staff whose motivation is research. They are not at all willing or skilled to be involved in web design and are too expensive to assign to routine technical work. The lack of appropriate personnel is often found in universities where little effort is put into developing elaborate courses on basic subjects. Rather, the pride of the alma mater is that well-known researchers contribute newest results in their teaching. In essence, content is much more important than presentation and the students accept this! There is no evidence that this situation will change in the future and any "Virtual University" has to take it into account.

2.5 Lecture Recording

The considerations of the last section led to the development of a method appropriate to a university teacher's "production process". The basic idea of *Electronic Lecture Recording* is to record a regular live lecture in order to obtain versatile digital material at negligible additional cost (Kandzia and Maass 2001).

Lecture recording allows for a wide range of flexibility. *Formats* depend on the expected user behaviour. If easy access over a browser is important streaming formats like Real or Flash can be used. On the other hand, if the provision of functions such as effective "fast forward" and "fast backward" is necessary (since difficult content requires intensive searching, scrolling and jumping within the recording by the learner) a so-called random access format like AOF (Müller and Ottmann 2001) is needed. The *teacher's style* (and the *available equipment*) range from a low-end variant (using photo-snapshots of a chalk blackboard together with audio and video), through PowerPoint (using previously prepared slides on an electronic whiteboard), to a highly interactive and lively style (featuring hand-written annotations on a graphical input device during the lecture). If appropriate, the presentation may contain computer animations and simulations.

In any case all data streams like audio, video, whiteboard activity, and animations are recorded automatically. They are stored on a server, and can be replayed in a synchronised way immediately after the live event. The recordings serve as an excellent basis for online courses. To that aim the recording is edited and exercises, self-assessments, or other supplementary material are added. The enhanced recording can now be integrated into an LMS.

When post-processing a lecture recording one can adopt an arbitrary quality in a spectrum between pure recordings without any editing and, for example, a Macromedia-WBT that comprises only short patches of thoroughly edited recordings. The producer can choose the level. Starting with no, or very simple, post-processing in the beginning, a provider can concentrate on tutoring and setting up the course management. Up-to-date content and interesting material for the student is much more important for success than appealing design. As resources allow, or if the course is repeated, material may be reworked and enhanced.

The main advantages of lecture recording are its cost effectiveness and its rapidity (which provides for very up-to-date material and swift updating when necessary). Also, due to the flexibility of lecture recording, the teacher's normal work is only slightly changed. The teacher is still preparing and holding lectures in their own personal style. He/she is not designing WBTs. Compared with a WBT the pedagogical potential of lecture recording may be restricted, but our experience is that students greatly appreciate lecture recordings providing they feature an interesting theme, a skilled teacher and appropriate support. The only serious drawback is that no fully convincing system for lecture recording is available on the market at present but several interesting products are emerging.

2.6 Tutored Assignments

In studies of the natural sciences, Computer Science, or Mathematics conceptualisation of a subject takes place during lectures. As a pedagogical counterpart tutored assignments are necessary and foster dialogue and construction. Characteristically, depending on the lecture course in which a student is enrolled, they will receive assignments to be solved alone or in a group of two or three persons. After a fixed amount of time, often a week, their solution is presented to a tutor for assessment. Usually tutors are advanced students responsible for ten to thirty students. The assessment is often part of the course accreditation. Afterwards the tutor discusses the solutions and other questions concerning the lecture with the student group.

This process, which is very typical in a university, can be transferred to web-based distance learning. In addition, such a system should include a report pool where all tutors of a course note general mistakes students have made and good solutions to the exercises. This tool also organises the important feedback between teacher and tutors. Statistical functions give information on whether the exercises have been too easy or too difficult. The statistical functions are much easier to implement and use in the virtual environment than in the paper-based setting. The students should have the possibility of comparing their personal marks with the average. Because exercise development is valuable work, a pool is needed in which all exercises are stored for reuse, together with appropriate metadata.

Surprisingly, although a lot of work has been undertaken on intelligent tutoring systems and adaptive exercises (Brusilovsky and Miller 2001), few systems have been designed to support the specific tasks and the communication features described above. In particularly the features of widespread commercially available LMS like Lotus Learning Space or WebCT are not sufficient. The FernUni Hagen (a German distance teaching university) was forced to optimize its paper and snail-mail based communication so it developed WebAssign and uses the system with thousands of students. Over thirty universities also use that open-source software. WebAssign is specifically designed for creating exercises in programming (Brunsmann et al 1999).

Another popular system is also called WebAssign and was developed by the North Carolina State University. It has similar features to the FernUni Hagen WebAssign package. One drawback is that the integration into an overall LMS system (e.g. access to user data) has to be done by the adopting institution.

At present no system on the market fully supports all the functions we have described. However, we think that in the near future, appropriate solutions will be integrated in LMS (some of our ideas will be implemented in the LMS 'CLIX' produced by im-c).

Tutored assignments over the web fit very comfortably in the everyday work of the alma mater. This is a necessary and logical completion of a lecture recording with which it shares similar advantages. Bringing both principles together enables a university to transfer a major part of its teaching and learning in the virtual world.

3. CONCLUSION

How can a traditional university open a virtual branch? Clearly, some basic functions like an LMS or facilities for videoconferences have to be installed and supported as a central service. Departments have to decide to what extent they want to "go virtual" - maybe a full masters' program in distance learning mode, selected courses, simple enhancement of classroom-based work or organisational services like enrolment over the web. In all cases basic issues have to be resolved before any implementation begins:

- Who are the users of the planned product?
- Is it a convincing offer for them?
- Are enough appropriate partners supporting the project?
- Are the partners able to finance the project both short and long term?

Clearing up these issues early on seems fundamental but surprisingly few projects are able to give satisfactory answers.

When implementing the offering it is a good idea to use low-expenditure scenarios like lecture recording, tutored assignments, or web-based seminars, which are closely related to traditional university work. Trying to change pedagogical, technical and personnel issues at the same time is not an effective approach.

The alma mater will not be swept away. Most of us will be glad about that. The approach favoured above will not cause a revolution in universities. Some will regret this. However, the approach offers a realistic chance for universities to build a programme of web-based distance courses with a scope broad enough to give students noticeably more flexibility, without losing the advantages of a campus-based education.

REFERENCES

Brunsmann, J., Homrighausen, A., Six, H-W. and Voss, J. (1999) Assignments in a Virtual University - The WebAssign System. In *Proceedings of the19th World Conference on Open Learning and Distance Education ICDE*, Vienna, Austria.

Brusilovsky, P. and Miller, P. (2001) Course Delivery Systems for the Virtual University. In *Access to Knowledge: New Information Technologies and the Emergence of the Virtual University*, F. T. Tschang and T. Della Senta (eds.), Elsevier, Amsterdam.

Encarnacao, J. L., Leidhold, W. and Reuter, A. (1999) Szenario: Die Universität im Jahre 2005. In *Zukunft gewinnen - Bildung erneuern*, R. Herzog (ed.), Goldmann, München.

Kandzia, P-T. and Ottmann, T. (2001) How real is the Virtual University in the Upper Rhine Valley? *International Journal on Engineering Education*, Vol. 17, No. 2.

Kandzia, P-T. and Maass, G. (2001) Course Production - Quick and Effective. *3rd NLT/NETTIES*, Fribourg, Switzerland.

Müller, R. and Ottmann, T. (forthcoming) Electronic Note-Taking, Systems, Problems, and their Use at Universities. In *Handbook on Information Technologies for Education & Training*, H. H. Adelsberger, B. Collis and J. M. Pawlowski (eds.), Springer.

Rumble, G. (2001) The Costs and Costing of Networked Learning. *Journal of Asynchronous Learning Networks* (JALN), Vol. 5, Issue 2.

Teacher training - the interplay of IT and society

Christian Görlich and Ludger Humbert
Studienseminar für das Lehramt für die Sekundarstufe II, Museumsstrasse 8, D-59065, Hamm, Germany

{goerlich|humbert}@semsek2.ham.nw.schule.de

Key words: Ethical/Ethics, Knowledge Management, Learning Environments, Teacher Training, Values

Abstract: This paper emphasizes the necessity of approaching the conference agenda with a more precise terminology and more practical and verifiable projects. It proposes that data related to current on-line learning projects could be made available for empirical research to provide theoretical grounding and facilitate reconstruction of sociological questions. The paper argues that purely scientific approaches are too narrow and limited in scope. It draws attention to ethical implications and supports Merton's view of the decisive role and significance of values to be observed. Focusing on the debate of computer supported collaborative learning (CSCL) the authors suggest that any research should consider in a more systematic and consistent way than before the individuals concerned, the constantly changing structure of social groups and the given social context. The authors build on the work of Holzkamp (who demanded a theory of learning based on the individual without ignoring the social context) and Foucault (who considered educational structures and institutions served to guarantee the efficiency of forms of learning and function as an element of control and social order). They stress that change in a CSCL environment may result in fears and anxieties that have to be given due consideration. The authors report on a current CSCL project in teacher training. Within the context of clearly defined academic research trainees explore and test new forms of cooperation and "knowledge management". This is effected by creating and working on virtual learning stations in different places and at different times. Preliminary findings indicate that educational bureaucracies and government offices need to adjust now and in the future.

The authors trust that more empirical data will become available for educational research focussing on the learning biography of adult learners.

1. IT - THE HISTORICAL AND SOCIOLOGICAL PERSPECTIVE

Using the word "new" in the context of new media and IT may suggest preconceived notions we are not aware of. It seems to suggest that not only the subject matter at hand (in this specific case computer studies/IT) is "new", but also the related problems. Surely this is only true in some respects - in other respects it requires modification. No-one should forget the historical dimension of current issues and controversies. There is no doubt that any debate today will benefit from consideration of such historical studies.

To a certain extent new technologies are means of production which interact in various ways with different social classes and élites. In this context our attention should be focused on the wealth of accumulated knowledge and acquired skills in any given society. This is one of today's most crucial assets. Against this background "the interplay of 'new' technologies and society" does have a comparatively long tradition of related problems.

Modern philosophers have found it increasingly difficult to refute the Marxist notion that socio-economic forces exercise a stranglehold on all cultural phenomena. Neo-Marxists have often simplified and reinforced these tenets. It may be superfluous to point out that Marx himself had a more differentiated concept of the process of (and consequently the means of) production.

In both sociology and cognitive science there have been a number of new approaches since the beginning of the 20th century. Representatives of these new schools of thinking are the German philosophers Max Scheler and Karl Mannheim. This school is widely known and recognised as the "traditional knowledge-based sociological approach". Without ignoring its historical merits this approach has been severely criticised in respect of the limitations of its questions and its simplistic answers.

The structural and functional approach has replaced these post-Marxist schools of thinking. Two factors have proved to be decisive - the influence of natural sciences (biology in particular) as well as the negative consequences of scientific discoveries. This approach has paid special attention to systemic rather than isolated causally related relationships. The work of Robert K. Merton is important in this context. For Merton science is more than just accumulated knowledge, far more than a variety of skills. Science is:

"an organized social activity of men and women who are concerned with extending man's body of empirical knowledge through the use of these techniques. The relationship among these people, guided by a set of shared norms, constitutes the social characteristics of science." (Merton 1972)

In recent decades all relevant research has been based on these assumptions. It is simply a matter of speculation if and when a third approach might be used to discuss issues relating to the interaction of IT and society - the sociological and historical approach. IT as such does not have a "history" that would merit and deserve historical studies. Thomas Kuhn has contributed significantly to this school of thought refuting the positivist implications he was concerned about in the structural-functional approach (Kuhn 1962).

This brief survey of sociological and historical schools of thinking has only one objective - to remind us of the fact that the problems we are facing today are not "new" and therefore it may be both useful and worthwhile to refer to these debates.

The reference to Marx serves a second purpose. It reminds us that any debate on the interplay of society and IT is not just concerned with a cognitive interest but genuine emancipation. More is at stake than the interrelationship of technology and society. It is always about power, rule, property and even human suffering and pain.

The objective for our target group should therefore be rephrased as follows:

What exactly is the potential of IT to support the process of Enlightenment? How can this specific Euro-centric approach be reconciled to the global context of IT and the current discussions about a culture clash? Where are the potential stumbling blocks?

Merton has drawn our attention to the issue of the prevalent code of values within this process. In particular he refers to notions like "universalism"," communism", "philanthropism" and "organised scepticism". These notions are the tip of the iceberg of a complex system of values. They need to be discussed against the background of, for example, historical changes, the contrast between norms and reality, the issues of internalisation and educational measures as well as in terms of sanctions for deviant behaviour. It is most informative and telling to conduct research concerned with "the debate on priorities". Here you will discover how significant, and at the same time how controversial, the question of copyright and patents has been and still is in the history of modern science. These concepts have undergone a fundamental change.

In this context Merton emphasised the objective of medieval artists to remain anonymous. In contrast to these notions another aspect of the very same discussion emphasises the ever-present function of norms like copyright and patents. So any seemingly irrational pattern of individual behaviour could be explained within a systemic context.

These days the issue of copyright and patents has assumed a new quality and urgency - even if you only consider it indirectly as a form of assessment. Two factors have contributed to this situation - the private use of IT hardware and software as well the increased numbers of those involved in CSCL. An international conference held at Humboldt University (Berlin) in 2001 chose as their agenda:

"Whose knowledge is it? Good to know. Links for an informed society" [http://www.bildung2010.de/gutzuwissen/].

Without doubt free access to IT resources and sources of information will ultimately be decisive for the quality of learning in the future. More differentiated research projects have been presented in the wake of criticism against traditional sociological concepts. Society is an abstract notion that the individual can only experience indirectly in the context of personal daily routines. This premise would enable us to refocus the issue of IT and society against the background of genuinely educational concerns and allow us to explore how the cognitive structures of individuals (pupils, students, trainees and teachers) and their groups (CSCL) interrelate in the context of IT and society.

We have already emphasised the need for an open historical approach. However we need to ask ourselves how valid Kuhn's concept really is. How useful are his terms such as "paradigm" and "anomaly" and his distinction of "pre, while and post-paradigmatic periods"? Are they simply a deluding metaphor or would they open the door for constructive and promising forms of research?

2. CSCL - RESPECTING THE NEEDS OF BOTH INDIVIDUAL AND SOCIETY

The public debate around the key phrase "CSCL" is moving towards consideration of the possibilities of transferring computer-assisted processes of co-operative work from the commercial world into education. Members of the latter sector are generally expected to accumulate and communicate knowledge synchronously on a local level, though in the first place they should also be enabled to do this asynchronously and in a distributive sense (Wessner et al 2000).

This debate has its origins in the criticism of www-based tele-learning because of its implied limitations in interactivity, individualisation and feedback. Hopes were expressed that improvement on those three aspects would result from the use of CSCL. At the same time it has been pointed out that CSCL is increasingly used in a social context to save resources of time and money.

The authors would like to stress that CSCL has a number of underlying limitations very similar to those of traditional group work. Not all topics are suitable for co-operative work and not every student is in favour of this method of learning. Still, without discussing all pedagogical implications these thoughts can be regarded as being of limited relevance.

In a similar way superior levels of social structures become the focus of attention simply because of their presence and without any clearly recognisable intentions in emancipation. Such a limitation represents a challenge to all reflective pedagogical thoughts.

In his substantial work Klaus Holzkamp (1995) has quite correctly criticised the neglect of the individual in our understanding of the process of learning. In view of currently practiced concepts of constructive criticism of learning, which see the student as an autopoetical system, Holzkamp's criticism may irritate and appear exaggerated. On the contrary: one is likely to accuse cognitive oriented constructive critics of ignoring those indisputably existing restraints of our society with all its limitations in their school of thinking. However, in spite of his plea for an individual oriented school of learning, Holzkamp does not ignore them. His analysis of the genealogy of school as an institution, based on Foucault (1977), explicitly includes the sociological dimension of learning. Still, "society" is a rather abstract term and might only just be experienced by the individual within its direct social surroundings.

We therefore propose a more differentiated view of this complex field of learning. The relatively young institution of learning in modern schools stems from the same ambition to act rationally that can be recognised in the layout of hospitals or prisons. The concentration of a certain amount of people in the same place at the same time probably has its origins in the ambition to act efficiently, but it will also always be closely related to the social function of control. It is obvious that research of local/polylocal and synchronous/diachronous forms of learning not only refer to organisational matters. They go deep down to the roots of society. Any change of this familiar basic order of our society might cause fear such as a loss of control; a kind of fear not unknown to conventional pedagogics. Acting pedagogically requires a clearly focussed target. Basically this way of thinking follows an old teleological idea developed in Aristotle's time.

The aim of any individual action can be placed in a hierarchical order of pragmatic actions - at the top of which we find the aim of "leading a happy life". Over the past centuries in German speaking countries the term "education" has taken over the function of the process of focussing on targets. A lot of effort has been expended during each period of time to infuse this term with an appropriate current meaning. So, what is its meaning in our age of new technologies? Kuhlen (1999) put forward a suggestion which offers a very useful basis for further discussion. His theory concerns two interrelated chains of terms. In the field of IT he uses the terms "data", "knowledge" and "information" and in the field of linguistics he refers to the commonly used semiotic levels "syntax", "semantics" and "pragmatics". Connecting these two fields we can simply say that in pragmatics we deal with the application of information, in syntax we deal with the inner and outer structure of data, and in knowledge we deal with the sense of it all, the truth of a matter and also with education. Naturally, Kuhlen is aware of all the limiting aspects implied when he talks of Enlightenment in education with special emphasis on the "autonomy of knowledge".

"Autonomy of knowledge is an aim possibly never to be achieved in our bourgeois society; being able to solve one's problems with one's own resources of knowledge. Today, autonomy of information is a realistic aim. Autonomy of information is not equivalent to having already gained full knowledge. The result of the process of collecting information by means of IT helps to create autonomy of information".

Admittedly, all this reasoning remains to be determined by the emotionalism of Enlightenment. Discussions of pedagogics with "post-modern knowledge" (Lyotard 1994) are largely confined to academic circles and have hardly influenced the training of student teachers.

3. THE STUDIENSEMINAR HAMM PROJECT

Against the background of the above we would like to highlight the project of the Studienseminar Hamm, an institution in which future teachers for schools of higher education are trained. This institution is typical of others in many federal states in Germany. After completing their training at university, the young teacher trainees spend two years of their second stage of training at schools and at the seminarium. The schools and the seminarium co-operate closely to continue the training on a practical and theoretical level. Because of the underlying inner structure - where several partners share the training and the participating schools are spread over a wide area - efficient forms of training and co-operation have had to be developed. Hamm started off by building up a network for electronic communication (e-mail, etc.). On this basis, and in co-operation with the University of Dortmund, so called virtual learning stations are being installed, enabling polylocal and asynchronical work to take place, complementing the conventional education in classrooms and in face-to-face seminars. Though still in its early stages, many interesting developments relevant to our topic "new technologies and society" have emerged from this project.

- The structural arrangement of learning stations leads to on-the-spot articulation of problems which arise in the trainees' direct surroundings (his/her school), which then can be resolved through immediate investigation. In this process the young teachers find it very helpful to be able to rely on the categories of "knowledge management", such as "generating/accumulating knowledge", "communicating knowledge", "presenting knowledge" and "using knowledge". Using these categories they are made aware of basic structures of problem solving processes. This in turn gives rise to hope that a transfer on other areas of work will take place.
- The solution of (among others) technical questions requires a flexibility that traditional bureaucracies can not achieve. Sooner or later this forces the rearrangement of traditional hierarchies into smaller, interrelated units within a network. A short term measure by well-informed school authorities in reaction to this development is to allow for as much freedom to act as possible.
- A clearly recognisable rectification and interrelation of the stages of teacher training can be observed. Working at learning stations requires certain IT qualifications and the contents of learning stations are made available for the third stage in teacher training.
- Virtual learning stations allow for individual as well as cooperative learning. From a didactical point of view an in-depth evaluation of the work at learning stations promises to give an insight into and improvement of the individual and cooperative learning processes.
- The dynamics of the developments of new technologies subversively influences conventional and traditional understanding of roles, i.e. the separation of trainer and trainee. At best trainers may become consultants, but occasionally they become students again.

Acknowledgement

Special thanks to Klaus-Peter Krause and Bernd Freyer for their assistance in translation!

REFERENCES

Foucault, M. (1977) *Überwachen und Strafen. Die Geburt des Gefängnisses.* Suhrkamp, Frankfurt am Main.

Holzkamp, K. (1995) *Lernen. Subjektwissenschaftliche Grundlegung.* Campus, Frankfurt am Main.

Kuhn, T. S. (1962) *The Structure of Scientific Revolutions.* University of Chicago, Chicago.

Kuhlen, R. (1999) *Die Konsequenzen von Informationsassistenten. Was bedeutet informationelle Autonomie oder wie kann Vertrauen in elektronische Dienste in offenen Informationsmärkten gesichert werden?* Suhrkamp, Frankfurt am Main.

Lyotard, J-F. (1994) *Das postmoderne Wissen: ein Bericht.* Passagen Verlag, Wien.

Merton, R. K. (1972) Die Priorität bei wissenschaftlichen Entdeckungen. Ein Kapitel der Wissenschaftssoziologie. In *Wissenschaftssoziologie, volume 1, Wissenschaftliche Entwicklung als sozialer Prozess*, P. Weingart (ed.), Athenäum, Frankfurt am Main.

Storer, N. W. (1972) Das soziale System der Wissenschaft. In *Wissenschaftssoziologie, volume 1, Wissenschaftliche Entwicklung als sozialer Prozess,* P. Weingart (ed.), Athenäum, Frankfurt am Main.

Wessner, M., Pfister, H-R. and Miao, Y. (2000) Umgebungen für computerunterstütztes kooperatives Lernen in der Schule. *Informatica didactica*1, No. 1. [http://didaktik.cs.uni-potsdam.de/InformaticaDidactica/Issue1/]

Author Index

Keyword Index